Volume 11

FORECASTING TECHNIQUES
FOR URBAN AND
REGIONAL PLANNING

FORECASTING TECHNIQUES FOR URBAN AND REGIONAL PLANNING

BRIAN G. FIELD AND BRYAN D. MACGREGOR

Routledge
Taylor & Francis Group

LONDON AND NEW YORK

First published in 1987 by Hutchinson Education

This edition first published in 2018
by Routledge
2 Park Square, Milton Park, Abingdon, Oxon OX14 4RN

and by Routledge
711 Third Avenue, New York, NY 10017

Routledge is an imprint of the Taylor & Francis Group, an informa business

British Library Cataloguing in Publication Data
A catalogue record for this book is available from the British Library

ISBN: 978-1-138-49611-8 (Set)
ISBN: 978-1-351-02214-9 (Set) (ebk)
ISBN: 978-1-138-48055-1 (Volume 11) (hbk)
ISBN: 978-1-138-48059-9 (Volume 11) (pbk)
ISBN: 978-1-351-06250-3 (Volume 11) (ebk)

Publisher's Note
The publisher has gone to great lengths to ensure the quality of this reprint but
points out that some imperfections in the original copies may be apparent.

Disclaimer
The publisher has made every effort to trace copyright holders and would welcome
correspondence from those they have been unable to trace.

Forecasting Techniques for Urban and Regional Planning

Brian G. Field
Department of Building and Estate Management,
National University of Singapore

Bryan D. MacGregor
Department of Town and Regional Planning,
University of Glasgow

Hutchinson

London   Melbourne   Sydney   Auckland   Johannesburg

Hutchinson Education

An imprint of Century Hutchinson Ltd
62–65 Chandos Place, London WC2N 4NW

Longwood Publishing Group
27 South Main Street, Wolfeboro,
New Hampshire 03894-2069

Century Hutchinson Australia Pty Ltd
PO Box 496, 16–22 Church Street, Hawthorn,
Victoria 3122, Australia

Century Hutchinson New Zealand Ltd
PO Box 40-086, Glenfield, Auckland 10, New Zealand

Century Hutchinson South Africa (Pty) Ltd
PO Box 337, Bergvlei, 2012 South Africa

First published 1987
© Brian G. Field and Bryan D. MacGregor 1987

Set in 10/12 pt VIP Times by
D P Media Ltd, Hitchin, Hertfordshire

Printed and bound in Great Britain by
Anchor Brendon, Essex

British Library Cataloguing in Publication Data
Field, Brian
  Forecasting techniques for urban and regional planning. — (The Built
  environment series)
  1. City planning 2. Forecasting
  I. Title   II. MacGregor, Bryan   III. Series
  711′.4   HT166

  ISBN 0-09-173101-1

Library of Congress Cataloging in Publication Data
Field, Brian G.
  Forecasting techniques for urban and regional planning.—(The Built
  environment series)
  Bibliography: p.
  1. City planning 2. Regional planning.
  3. Forecasting. I. MacGregor, Bryan D.   II. Title.
  III. Series.
  HT166.F48   1987   003′.2   87-2650

  ISBN 0-09-173101-1 (pbk.)

Contents

Preface

The original idea for this book came as the result of conversations we had when based at Cambridge. Prior to that we had both come to planning from numerate first disciplines and, in planning schools at opposite ends of Britain, had independently concluded that there was an obvious gap in the literature on this particular subject. On the one hand, several fine theoretical texts and monographs had been produced which, although extremely useful to academics, were often unnecessarily obfuscating to the less numerate reader assuming, as they did, a mathematical competency far beyond the capabilities of the average undergraduate or practitioner. At the other extreme, there were the more general 'literature review' on 'survey' texts which briefly introduced the reader to a number of techniques without explaining how these might be applied in practice. Once again, these were of only limited value — popular with the first-year undergraduate but of little use subsequently, and almost useless as far as practitioners were concerned.

The focus of this text is on those techniques which are commonly employed in making forecasts for subsequent use in plan-making and policy formulation. The intention is to provide a practical guide which offers students and practitioners a concise overview of the selected techniques supported by an appropriate collection of worked examples. The aim is not only to familiarize readers with particular techniques but, through the worked examples, to demonstrate how to apply such techniques in practice. Moreover we try to place the techniques in the context of policy formulation rather than to portray them as merely technical constructs. A further feature of our approach is that we have tried throughout to adopt a critical approach and have sought to identify the practical limitations of the techniques in question. In this way we hope that they will be used more appropriately and with better understanding of their shortcomings.

There are two important consequences of the approach as outlined above. First, we cannot be comprehensive. Since the focus is on forecasting techniques in planning analysis, large areas, such as plan generation, evaluation and monitoring are left untouched. These other aspects, while important, must be the subject of other texts. It is inevitable that we will be criticized for missing out this or that technique but, in our view, we have covered, in a reasonably comprehensive manner, a clear sub-set of analytical techniques in common use in planning: others may disagree with this judgement. Second, we have been unable to deal with the relationships between theory and technique in other than a cursory manner, but readers should not allow this apparently perfunctory treatment of theoretical issues to negate their significance. Needless to say, we would consider a clear understanding of theory to be fundamental to an accurate appreciation of the techniques and their limitations. However, detailed theoretical issues are probably best covered elsewhere.

1 Introduction

1.1 Summary

This book is an introduction to the various analytical techniques related to forecasting which have been developed and applied by urban and regional scientists in practical planning situations. It is not the intention to provide a compendium of the theoretical and methodological niceties of the numerous techniques bequeathed us by the quantitative revolution, but rather to focus on some of the more important methods and to demonstrate their utility in planning analysis: in other words, the approach is practical. The aim is not only to familiarize readers with selected techniques but, by adopting a step-by-step approach of worked examples, to give them the confidence to employ these in making forecasts.

Experience suggests that planning students and, indeed, some practitioners, find considerable difficulty in dealing with numerate methods, and this prevents a full appreciation of the strengths and limitations of the tools of quantitative analysis that can be brought to bear in the planning process. Recognizing this problem, the book is deliberately structured to assist those with a relatively limited numerate background. Clearly it is impossible to deal with analytical techniques without some recourse to the language and methods of mathematics and statistics but, by presenting each technique both verbally and mathematically, and by illustrating its operating characteristics with detailed but simple practical examples, the book tries to avoid obscuring the reader's understanding with unnecessary technicalities.

1.2 The planning process

In UK planning practice, the growing importance of analytical and predictive methods is, to a large extent, the product of attempts following the advice of the Planning Advisory Group (PAG, 1965) to create a more rational approach to plan-making. Batey and Breheny (1978) have high-lighted two ways in which planning has, as a result, become more systematic. First, and perhaps most important, was the rejection of Patrick Geddes' (1915) rather disjointed procedure of survey → analysis → plan, which had provided the methodological underpinning for land-use planning for almost 50 years, and its replacement and transformation by a more logical approach drawing on decision theory and systems analysis. The result was a shift in the focus of planning, from the production of a single product to delineate an ideal end state to be achieved at some future date (*master planning*), to more of a continuous process and programme-orientated activity in which it is assumed that objectives are clearly defined and that planned actions follow sequentially from decisions or policy. Evaluation, which is central to such a process and occurs both implicitly and explicitly, measures the extent to which actions taken are directed towards the achievement of policy and objectives. Planning is no longer perceived as a random set of activities brought together to achieve some blueprint for the future; rather the inter-connectedness of decision areas is explicitly recognized in a cyclical process which has no definite beginning or end, but is designed to enable planners to address new problems as and when they arise. McLoughlin (1965) summar-ized this 'systems approach' to planning as follows:

> Planning is not centrally concerned with the design of the artefacts, but with a continuing process that begins with the identification of social goals and the attempt to realise these through the guidance of change in the environment. At all times the system will be monitored to show the effects of recent decisions and how these relate to the course being steered. This process may be com-pared to that encountered in the control mechanisms of living organisms, part of the subject matter of cybernetics.

There are numerous formulations of this process (see, for example, Field, 1984), but certain common elements recur and a consensus view might be represented by the following stages:

1   the definition of objectives in relation to more general goals, or in explicitly problem-solving exercises, the identification of actual problems and issues;
2   the generation of strategies to achieve goals and objectives, and the formulation of policies to address particular problems;
3   the testing and evaluation of strategies and policy packages;
4   implementation and monitoring.

The overall process is illustrated schematically in Fig. 1.1, which demon-strates the links between the various stages (more detailed elaborations of the process show more lucidly the full cycle and how, through feedback

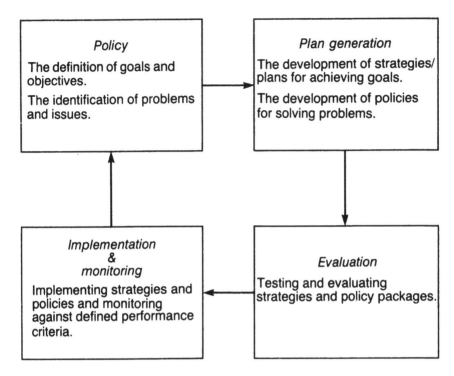

Figure 1.1 *The cyclical planning process.*

loops, decisions can be modified when the consequences of policies are exposed). Although a number of variations of this basic approach have emerged over the years — principally the result of commentators highlighting the disparity between the requirements of the 'pure' systems model and the capacity of decision makers — there has been a general acceptance of the need for more rationality. But it is not just the transformation of the planning process which has promoted the rational decision-making approach; planning has become more systematic in a second important respect, namely it is now based on the use of more formal replicable methods of analysis at each stage within the overall process, as compared with the informal 'rules of thumb' which characterized planning practice in the immediate post-war period. In this respect, the planners' lexicon or 'tool kit' has changed quite markedly as the techniques of quantitative analysis have been brought to bear in the study of the structural relationships among the variables within the system.

In this text, the emphasis is on those methods of analysis which improve our understanding and provide quantitative assessments of the various components within the planning process, with an emphasis on forecasts.

Methods concerned with the broader process such as plan generation, and evaluation and monitoring, receive little attention, although it should be pointed out that their use has been one of the important consequences of the quantitative revolution. This has necessarily led to the categorization of techniques by subject — for example, population, housing and employment — paralleling the general use of separate 'topic areas' for analysis and policy formulation in structure and local plans which, indeed, closely follows the advice of the *Development Plans Manual* (Ministry of Housing and Local Government (MHLG), 1970). A number of commentators have criticized this preoccupation with discrete subject area analysis, which is in many ways far from systematic. However, an important objective of the book is simplicity, and this is best achieved by the arrangement into chapters covering separate topics, while clearly indicating the extent of interconnectedness. Nevertheless, an attempt is made to integrate these partial subject analyses, through their crucial supply and demand linkages, in Chapter 8.

1.3 Models in planning

The tools of planning analysis include an extensive range of concepts and techniques which have been developed to understand and articulate the nature of cities, and to predict the consequences of change. These range from the very basic methods of descriptive statistics to the construction of more complex mathematical models. Since concern here is primarily with the inferential and, particularly, predictive properties of these techniques, the book's main concern is with those methods which can be utilized to provide forecasts in planning. Techniques of this kind are invariably based on mathematical models.

A model is simply a way of representing reality in which real-world objects and relationships are expressed, either physically or in the abstract, in some way relevant to their characteristics. Their value lies in improving understanding of reality where it is not possible to construct an experiment in the real world. Physical models can be created by replicating the original at a different scale, for example an architectural model of a proposed development, or by analogy, for example a map or plan. By contrast, abstract models represent the relevant characteristics of the original in a language, which might be expressed verbally using words and syntax or mathematically using symbols and equations. Mathematical models lend themselves to easy manipulation by avoiding many of the problems of the other variants, such as the ambiguity that often characterizes verbal conceptual models or the inflexibility of the physical model. This facilitates their use in the making of forecasts. But perhaps, most important of all, they provide the opportunity for computation.

1.4 Model characteristics

An urban region is a highly complex system of interconnected activities which, for ease of understanding, can be divided into subsystems. When building a planning model it is necessary to try to understand and describe the mechanisms which govern the behaviour of a given subsystem, and then translate what is often a verbal description into the language of mathematics. But the real-world subsystems are themselves very complex, so in order to make models workable, they are going to be considerably simplified representations of our phenomenon of interest. Nevertheless, if properly constructed, they can provide a good structure for problem solving and evaluating planning policies generally. The formulation of a typical planning model can be considered in five stages (after Lee, 1973): 'selection of the variables to be included; choice of appropriate level of aggregation and categorization; decisions about the treatment of time; specification; and calibration.'

1.4.1 The selection of variables

Almost all mathematical models assume that changes in the value of the real-world variable under study (the *dependent variable*) are in some way dependent on a combination of changes in other real-world variables (the *independent variables*). Since the model is to be a simplified representation of reality, an important question facing the model builder is, therefore, deciding which variables to include. In the first instance he/she is influenced by the definition of the problem at hand, but the determining factor is the strength of the casual relationships between the dependent and independent variables. All variables which display a strong casual relationship should be included in model formulations. For example, assume that the dependent variable is A, and that B, C and D have been identified as the important independent variables. Using mathematical notation, this combination can now be written as follows:

$$A = f(B, C, D) \tag{1.1}$$

Equation (1.1) is a simple functional relationship which states that the value of A depends on, or is a function (f) of, the values of B, C and D, that is, A will change if there is a change in the value(s) of B and/or C and/or D. (See the Appendix for a fuller discussion of mathematical notation.)

1.4.2 Level of aggregation and method of categorization

Having decided which variables to include, decisions need to be made about how these should be categorized and the appropriate level of aggregation. The choice of a given method of categorization is determined by the purpose of the model, for example, in a population study the classification might well be according to age and sex. As to level of aggregation, this differs between models, but it is rare to develop micro-models that deal with individual behaviour. In a transport or shopping study, for example, the models deal with aggregate behaviour patterns within and between zones, so if the concern is with the number of shopping trips made by residents from a predominantly housing area, say zone i, to another predominantly shopping area, say zone j, then the simple equation (1.1) might now take the following form:

$$A_{ij} = f(B_i, C_j, D_{ij}) \qquad (1.2)$$

where A_{ij} = the number of trips from i to j
  B_i = the resident population in zone i
  C_j = the number of shops or amount of retail floorspace in zone j
  D_{ij} = distance between i and j.

1.4.3 The treatment of time

The treatment of time in a model relates both to the theory on which the model is built and to the facility with which it can be used to make forecasts. The latter is an essentially operational consideration and depends upon the time period for which the model is to be used, that is the time horizons of the plan. The first aspect is more problematic, however, and relates to the way in which the passage of time is accounted for in the model itself. Almost all practical planning models are static or comparative static equilibrium models, that is they deal with discrete observations in time and operate as though these represent equilibrium states of the subsystem, which can then be easily extrapolated to provide forecasts. In this way they fail to acknowledge that urban change is a continuous and dynamic process. However, the theories which have been developed to explain the dynamic workings of the urban system are far from adequate and considerable problems, certainly beyond the scope of an introductory text like this, are encountered in trying to build even simple dynamic urban models.

1.4.4 The specification of the model

For equation (1.2) to be useful for prediction it is necessary to translate the general hypothesis, that the value of A is dependent on the values of B, C and D, into a more formal mathematical relationship that specifies the nature and degree of dependence. The nature of dependence may be direct, that is an increase in B results in an increase in A, or inverse, that is a decrease in B results in an increase in A. Given the simple example of equation (1.2), observation suggests that the nature of the relationship is such that the number of trips, A_{ij}, between i and j depends directly on the size of the resident population in i, B_i, and on the attractiveness of j as a destination, C_j, and inversely with the distance between i and j. This can be written as:

$$A_{ij} = f\left(B_i, C_i, \frac{1}{D_{ij}}\right) \tag{1.3}$$

Equation (1.3) can now be refined in mathematical terms by the inclusion of various constants or parameters which add dimensions to the model's relationships and specify the degree of dependence, that is, they tell by how much one variable changes in relation to others. This is not always a straightforward process but, since the operational model being developed is based on the familiar gravitational concept (which is discussed at length in Chapter 5), it can be written as:

$$A_{ij} = \frac{k \, B_i \, C_j}{D_{ij}^d} \tag{1.4}$$

where A_{ij} = the number of trips between i and j
B_i = the resident population in zone i
C_j = the number of shops or amount of retail floorspace in zone j
D_{ij} = the distance between i and j
d = a constant
k = a balancing factor

1.4.5 Calibration of the model
Calibration is the process of finding numerical values for the constants in order to establish the precise degree of dependence between variables. This is normally achieved by 'fitting' the model to an observed situation in which the values of all variables, both dependent and independent, are known. The values of the constant and the balancing factor can now be derived, and the calibrated model might take the following form:

$$A_{ij} = k_i \frac{B_i \, C_j}{D_{ij}^2} \tag{1.5}$$

The model's ability to reproduce the characteristics and behaviour of the real world must now be tested by operating it in past or other present situations, checking the results against actual outcomes to see if any adjustments in model design are necessary. The various stages in the model-building process are illustrated in Fig. 1.2 (overleaf).

Predictive models of the type just described are what analysts call *deterministic*, because the output of the model is completely determined by the set of equations which describe the subsystem under study. Almost all the methods that will be described in this text are deterministic. By contrast planning analysts have, from time to time, developed *stochastic* models which incorporate a probabilistic component into the mathematical relationships which describe the subsystem. However, these are still of a largely

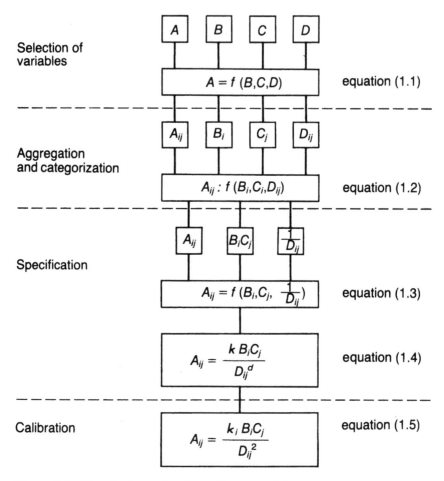

Figure 1.2 *Developing a simple predictive model.*

theoretical and experimental nature and are only touched upon, in view of our concern with more practical and operational techniques.

1.5 Forecasting

Planning is a process of analysis and action which is necessarily about the future. It involves intervention to manipulate procedures or activities in order to achieve goals. Forecasting is clearly crucial to such a process. The methods of analysis discussed thus far, and particularly the kind of planning models described above, not only improve our understanding of the various components within the spatial system but, most importantly, enable predictions to be made of future changes and the consequences of policy prescriptions. It is essential, therefore, that there is some measure of confidence in the forecasts generated by the methods of analysis and this 'validation' is usually achieved by measuring performance against three important criteria.

With any mathematical model, first and foremost, concern must be with the quality of the predictions that it makes. In other words, concern is with its *accuracy* as a forecasting tool, and this can be easily tested using well-established statistical procedures. But accuracy is not enough; concern should also be with the model's *validity*. It is quite possible for a model which does not properly describe or replicate the subsystem under study to produce very accurate results. However, when the assumptions behind a model are inadequate and it fails to establish correctly the underlying cause-and-effect relationships between the variables, apparently good predictions can be extremely misleading and result in the promotion of very mistaken policy proposals. Linked to the issue of validity is the question of *constancy*. A valid model which accurately describes the system components and adequately specifies their interrelationships will clearly be accurate, but only if the relationships which exist at present can be expected to remain over time. Having satisfied the requirements of accuracy, validity and constancy, and assuming that adequate data is available, in particular reliable estimates of future values of the key independent variables, the model can then be used to make predictions with some confidence.

Predictions normally follow one of two forms. On the one hand there is the 'one-off' prediction which estimates the value of the variable under consideration at some end date, say 20 years hence, in which case the model output measures the assumed equilibrium at that time. The forecast in question can take the form of a point estimate, that is a single value, or a range which defines the limits within which the actual value is assumed to fall. On the other hand, analysts frequently perform 'recursive' predictions. These involve breaking down the projection period into smaller intervals and forecasting for each interval in turn, with output from the first such prediction being used as additional input information for the next, and so on. For a 20-year forecasting period, therefore, four such iterations might be performed. Recursive predictions of this kind have certain advantages because, although strictly speaking still exercises in comparative statics, they do, nevertheless, introduce a dynamic component into the forecasting process.

1.6 Contents and structure

For reasons already discussed, the book is organized in chapters which are differentiated by subject area. An important objective is simplicity, and this is reflected in the structure of these topic chapters which include both descriptive/explanatory and analytical/predictive aspects of the techniques under study. Although the text forms an integrated whole, each chapter can, nevertheless, be read independently and in most cases only a rudimentary knowledge of mathematics and statistics is required. However, as pointed out in the introduction to this chapter, any serious consideration of the quantitative methods used in planning analysis cannot avoid some contact with the language and symbols of mathematics and statistics. As a result, many other texts on techniques begin with an introductory section covering what are often referred to as 'mathematical preliminaries'. We have preferred to move straight to the main body of the subject by concentrating on activity forecasts for the key topic areas, and have relegated these mathematical/statistical prerequisites to an appendix. The latter has been kept as simple as possible and only addresses those concepts which it is necessary to understand in the discussion of the various methods covered in the text.

2 Population

2.1 Introduction

Assumptions concerning population are behind most important planning policies. The predicted population level is an important input into forecasts of housing demand and thus housing land requirements. It is also important in creating local demand for goods and services, and thus affects the level of local economic activity. More generally, population assumptions underlie investment decisions on schools, hospitals, roads, recreational facilities, and power and water supplies. Clearly, as many of these facilities are restricted to a particular age group, not only are aggregate population forecasts required but also disaggregations according to age, sex, occupation, etc.

For planners it is important to understand the value, limitations and accuracy of the various techniques used in making population forecasts and, in some instances, planners should be able to apply these techniques in practice.

In almost all cases the interest is in the *normal resident population*, that is, excluding persons such as tourists and armed forces personnel. The latter are often important but are typically treated separately and are dependent on Central Government policy.

Forecasting involves an examination of past data and the determination of the relationship between an appropriate *independent* variable or variables and the *dependent* variable, in this case population. Information on past levels, composition, and spatial distribution of population is generally available and of good quality, primarily from the ten-yearly census and the Registrar General's annual reports. This is in marked contrast to some of the other data required by planners for forecasting. Forecasts are made for different sizes and types of area — sometimes for the total population, while in other cases for selected components. Obviously the technique used will depend on the type of area, the timescale and the required level of detail, although available data, required level of accuracy, and staff resources are also important factors. Notwithstanding that the complexity of the techniques varies substantially, most use time as the independent variable. Population growth is assumed to follow a predetermined pattern through time and, in such cases, the most important aspect is establishing this pattern. In some cases, however, a variety of economic and housing variables are used. For these it is necessary not only to establish the relationship between the independent and dependent variables (the model), but also to *forecast* values for the independent variables which are then used in the model to produce the population forecast. Unfortunately, information for these variables is less available and of poorer quality than for population.

2.2 Aggregate approaches

These *aggregate* methods are also known as *simple* or *macro* approaches. They are used to forecast *total* or *aggregate* population, usually for large areas. Four types are discussed below:

1   trend-line methods;
2   comparative method;
3   ratio methods;
4   multiple regression method.

2.2.1 Trend line methods

These methods take time as the *independent* variable, that is, they assume that population growth follows a set pattern. They may be classified according to the assumed pattern: (a) linear trend; (b) geometric/exponential trend; (c) modified exponential trend; (d) logistic curve trend. In each case the past population data is analysed to determine the trend, and the trend is then *extrapolated* to some future date to provide the forecast.

(a)   Linear trend

The simplest case is a linear trend, that is, the increase in equal time periods is constant. The data shown in Table 2.1 is plotted in Fig. 2.1. It is clear that

Table 2.1   *Linear trend*

Year	Population
1951	100 000
1961	110 000
1971	120 000
1981	130 000
1991	?

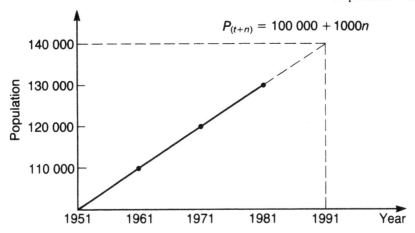

Figure 2.1 *Perfect linear trend.*

there is a linear trend, and a trend line can be drawn through the points. To forecast the population in 1991 using the *graphical* method the trend line is extrapolated and the figure for 1991 is read from the y axis. The answer is 140 000. It can be seen obviously from the table that population has been increasing by 10 000 each ten years and so will reach 140 000 by 1991 if *the same trend* continues.

The trend may be expressed mathematically by an equation for a straight line (see Appendix A.7). The equation for such a line is:

$$y = c + mx \tag{2.1}$$

where x is the independent variable
       y is the dependent variable
       c is a constant
       m is the gradient, which is constant.

In this case the equation may be written as:

$$P_{(t + n)} = P_{(t)} + mn \tag{2.2}$$

where n is the number of years between t and t + n
       $P_{(t + n)}$ is the population to be forecast at time t + n
       $P_{(t)}$ is the population at time t — the base year, y = 1951
       m is the gradient.

Using the method shown in Appendix A.7, the equation can be shown to be:

$$P_{(t + n)} = 100\ 000 + 1000n \tag{2.3}$$

For 1991, $n = 40$ (1991 – 1951) and so substituting gives $P_{(1991)} = 140\,000$.

In anything other than very simple examples, the mathematical method is quicker and more accurate than drawing a trend line and extrapolating it graphically.

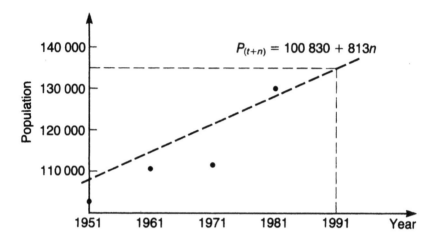

Figure 2.2   *Imperfect linear trend.*

In reality it is extremely unlikely that the data would fit a straight line so perfectly (see Fig. 2.2). In this case there are three possible ways to calculate the 'best' trend line.

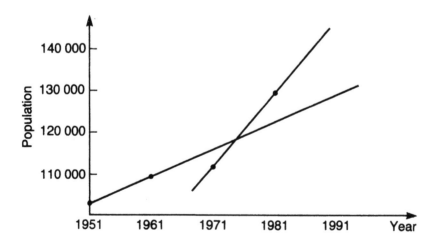

Figure 2.3   *Different lines from the same data.*

1   Using either the graphical or mathematical methods calculate the trend line between any two points. The problem arises that it is likely that no two pairs of points will produce the same line (see Fig. 2.3).

2   Draw the trend line by 'eye'. This need not pass through any of the points. While it is usually a better method than (1) above, it is possible that each forecaster will draw a different line and so produce different forecasts.

3   Calculate a regression line by the *least squares* method (see Appendix A.8). This ensures that all forecasters produce the same trend line and forecast.

Table 2.2   *Calculating the trend line*

Year	Population
1951	102 500
1961	109 900
1971	110 200
1981	129 500
1991	?

From the data given in Table 2.2 readers should calculate the trend line using each of these methods and compare their results. Note that the calculated lines in (2) and (3) above need not pass through any of the points and so the equation is:

$$P_{(t + n)} = A + mn \qquad (2.4)$$

where $P_{(t + n)}$, m and n are as before
A is a constant.

At this stage it is important that readers fully understand the linear trend method. The following trend methods, although mathematically more complex, involve the same reasoning and similar steps.

(b)   Geometric/exponential trend
In the preceding method it was assumed that population growth in equal time periods was constant, regardless of the total population. A more realistic method is to assume that the size of the increment in population is related to the size of the population. This method assumes that the ratio between the increment in population and the total population is *constant*, but the increment is increasing. This may also be stated as: for any given time interval n, the ratio of the population size at the end of the interval to that at the beginning is constant, or mathematically:

cont.

$$\frac{P_{(t + n)}}{P_{(t)}} = K \qquad \text{for all } n, \tag{2.5}$$

where $P_{(t + n)}$, $P_{(t)}$, n are as before
         K is a constant.

In this case it is difficult to sketch trend lines and the mathematical method is to be preferred. The data shown in Table 2.3 is plotted in Fig. 2.4.

Table 2.3 *Exponential trend*

Year	Population	Ratio	Growth rate
1951	100 000	1.1	0.1
1961	110 000	1.1	0.1
1971	121 000	1.1	0.1
1981	133 100	1.1	0.1
1991	?		

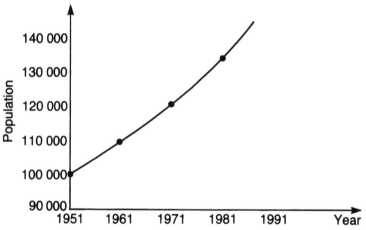

Figure 2.4 *Exponential trend.*

From the data it is calculated that population increases by one tenth or 10% *each ten years*, so the forecast for 1991 may be calculated from (2.5):

$$\frac{P_{(t + n)}}{P_{(t)}} = K \tag{2.5}$$

This may be rewritten as:

$$P_{(t + n)} = K \, P_{(t)} \tag{2.6}$$

In this case $n = 10$ and $K = 1.1$, and substitution gives

$$P_{(1991)} = 1.1P_{(1981)}$$
$$= 1.1 \times 133\ 100$$
$$= 146\ 410$$

More generally the formula for *geometric* growth may be expressed as:

$$P_{(t + n)} = (1 + r)^n P_{(t)} \qquad (2.7)$$

where $P_{(t)}$ and $P_{(t + n)}$ are as before
      n is the number of years between t and $t + n$
      r is the *annual* growth rate.

(Note that this produces a series of points, *not* a line.)

The growth rate r may be calculated by taking logarithms of both sides of equation (2.7) and expressing the equation in terms of r (see Appendix A.10). Thus:

$$r = \text{antilog}\left[\frac{\log P_{(t + n)} - \log P_{(t)}}{n}\right] - 1 \qquad (2.8)$$

Note that in the example shown in Table 2.3 the growth rate for *ten years* was 0.1. This is *not* equivalent to an *annual* rate of 0.01, but to an annual rate of 0.0096. (The reader should confirm this and read Appendices A.10 and A.11.) Substituting this annual growth rate in equation (2.7) gives:

$$P_{(1991)} = (1 + 0.0096)^{10}P_{(1981)}$$
$$= 1.1 \times 133\ 100$$
$$= 146\ 410, \quad \text{as before}$$

By this method a population level can be forecasted. However, as with linear trends it is unlikely that the data will fit the trend line exactly and so it is necessary to calculate a 'best fit'. In this case a straight line can be fitted by applying a *logarithmic transform* (see Appendices A.10 and A.11), and taking $\log P_{(t)}$ rather than $P_{(t)}$ as the dependent variable. As before it is best to calculate a regression line by the *least squares* method. The calculated line will have the form

$$\log P_{(t + n)} = B + n \log (1 + r) \qquad (2.9)$$

where $P_{(t + n)}$, r and n are as before
      B is a constant.

Table 2.4 *Calculating trend line using logarithmic transformation*

Year	Population	$\log_e P_{(t)}$
1951	98 000	11.49
1961	112 000	11.63
1971	125 000	11.74
1981	140 000	11.85
1991	?	?

Calculated regression line is

$$\log_e P_{(t+n)} = 11.50 + n \times 0.012$$
so
$$\log_e P_{(1991)} = 11.50 + 40 \times 0.012$$
$$= 11.98$$
so
$$P_{(1991)} = 159\,532.$$

If antilogarithms are taken of both sides equation (2.9) becomes:

$$P_{(t+n)} = (1+r)^n C \tag{2.10}$$

where C is a constant $=$ antilogarithm of B.

From the data given in Table 2.4 readers should calculate the trend line and use it to forecast the population in 1991.

For mathematical reasons which need not bother the reader it is possible and convenient to express the pattern of growth as:

$$P_{(t+n)} = Ce^{an} \tag{2.11}$$

where $P_{(t+n)}$, n, C are as before
e is the natural or exponential base
α is the growth rate.

(This equation produces a trend line rather than a series of points.)
(*Note:* $\alpha = \log_e(1+r)$)

(c) Modified exponential trend

Both of the above methods allow for an infinitely large population — population growth continues indefinitely. This is clearly an unacceptable assumption for many cases. A modified version of the exponential curve in which the absolute increases are decreasing in size and there is a maximum population is often more realistic. In Fig. 2.5 the maximum population is K and the ratio

$$\frac{K - P_{(t+n)}}{K - P_{(t)}} \quad \text{is constant for all } n.$$

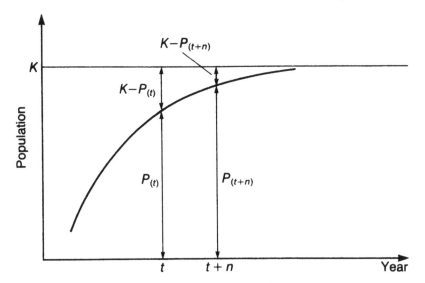

Note: it can be shown that $\log(K - P_{(t+n)}) = \log(K - P_{(t)}) + nv$

$$\text{where } v = \frac{1}{m} \sum_{t=2}^{n} \frac{K - P_{(t)}}{K - P_{(t-1)}}$$

Figure 2.5 *Population approaching a maximum.*

As before a trend line may be fitted. However, as the mathematics is more complex a worked example is not shown here.

(d) Logistic curve trend

Using the modified exponential is not without problems, because the curve allows for negative population, which is an absurdity. In this case the interest is only in the upper section of the curve. Similarly with the exponential, interest is in the lower section. The two approaches may be combined to produce an S-shaped curve (see Fig. 2.6). Growth begins slowly, but the size of each successive increase increases until it reaches a maximum and then continues to decrease. Curves of this form are called logistic curves. The mathematics involved in fitting trend lines to data of this form is too complex to explain here. It has been suggested that such curves best describe the patterns of growth in a confined area. However, in practice it has been shown that there are substantial fluctuations from the trends.

The above methods have all been based on the simple and simplistic assumption that population growth follows a fixed and therefore predictable pattern. In these models time is the *independent* variable. The following sections now examine methods which use other independent variables.

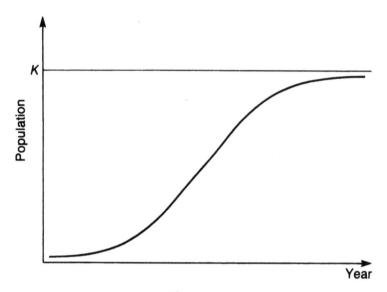

Note: the equation is $P_{(t+n)} = Ka^{b^n}$
which can be transformed to $\log P_{(t+n)} = \log K + (\log a)b^n$ (modified exponential)
or $\log (\log P_{(t+n)} - \log K) = \log (\log a) + nb$ (linear)

Figure 2.6   *The logistic curve.*

2.2.2 The comparative method

The techniques discussed previously have assumed that the future growth of
an area is predictable on the basis of knowledge of past growth, from which a
trend line may be established. By contrast the comparative method assumes
that the future growth of the *study* area will follow the pattern of another
older area called the *control* or *pattern area*. In Fig. 2.7 curve P represents
the pattern area and curve S represents the study area. In this case there is a
lagging, the distance AA′, of forty years. To forecast the population of area
S in 1991 it is only necessary to establish the population of area P in 1951.

In practice this is not a particularly useful technique. The fundamental
problem is choosing an appropriate pattern area. This area needs to have
experienced similar economic, social and political forces to those operating
on the study area. Such comparison requires a complex analysis which is
usually impracticable. In any case, the underlying economic development
theory assumptions of independent growth are no longer widely accepted.
The comparative method amounts to little more than an examination of
patterns of population growth rather than the underlying causes, and so is
similar to trend-line methods. It has been applied with some success to
suburbs in the same metropolitan area (Isard, 1960).

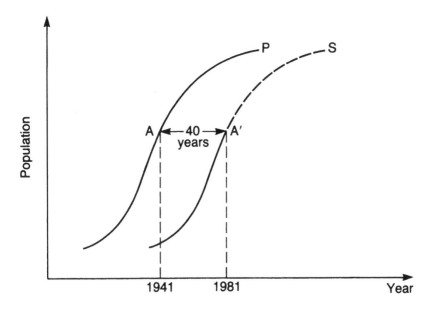

Figure 2.7 *The comparative method.*

2.2.3 Ratio methods
The term *ratio methods* is applied to a variety of different techniques which compare the study area's population with a *ratio variable* at the *same* time. The ratio variable is forecasted and then multiplied by a *ratio* to produce the forecast for the study area population. These methods may be conveniently categorized according to the ratio variables they employ:

(a) Area ratio method
(b) Component ratio method
(c) Symptomatic ratio method

(a) Area ratio method
In this variant of the ratio method the study area is compared with a pattern area of which it is part. For example, a region of a nation may be compared with the nation of which it is part. If the study and pattern area have been experiencing similar economic, social and political forces which have influenced population change, then it is hypothesized that there will be a relationship between growth in both areas. Table 2.5 shows population figures for a pattern and a study area. They are graphed in Fig. 2.8 (overleaf). The assumption is that:

$$P^s_{(t)} = R \ P^p_{(t)} \tag{2.12}$$

cont.

Table 2.5   *The ratio method: constant ratio*

Year	Population of pattern area (millions)	Population of study area	Ratio of study area to pattern area
1951	1.0	100 000	0.10
1961	1.1	110 000	0.10
1971	1.2	120 000	0.10
1981	1.3	130 000	0.10
1991	1.4 (forecast)	140 000 (forecast)	0.10 (forecast)

where $P^s_{(t)}$ is the population of the study area at time t
$P^P_{(t)}$ is the population of the study pattern area at time t
R is the ratio.

For the data shown $R = 0.10$, and so if the forecast for the pattern area for 1991 is 1.4 million then the forecast for the study area is 140 000.

It is unlikely that the ratio will be exactly the same for each time period. The easiest way to overcome this is to make the forecast based on an average ratio. Such a case is shown in Table 2.6. It is also possible to fit a least squares regression line to the ratios, as in Section 2.2.1, and to use this as the basis for forecasting.

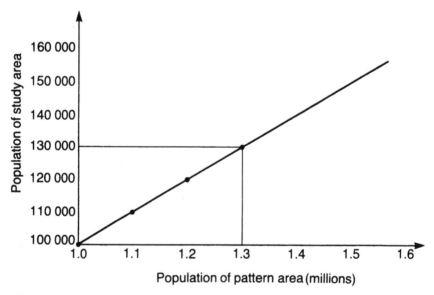

Figure 2.8   *Relating the study area to the pattern area.*

Table 2.6   *The ratio method: average ratio*

Year	Population of pattern area	Population of study area	Ratio of study area to pattern area
1951	1 025 000	103 525	0.101
1961	1 099 000	107 702	0.098
1971	1 102 000	109 098	0.099
1981	1 295 000	132 090	0.102
1991	1 333 500 (forecast)	133 350 (forecast)	0.100 (forecast) (= average)

There is no reason why the ratio should be constant. Indeed forecasts may be made of future values of the ratio using the trend-line analyses shown in Section 2.2.1 or by a more subjective analysis which considers likely influences. More elaborate methods, such as those shown in Section 2.2.4, could be used but these are best employed as independent techniques and not as part of this technique. Table 2.7 (overleaf) and Fig. 2.9 show the relationship between the populations of the Highlands and of Scotland as a whole. They illustrate the dangers of the technique. An understanding of the changes in the ratio requires an understanding of the underlying economic and social forces.

If the method is being used to forecast the populations of all of the areas which comprise the pattern area it is necessary to ensure that the sum of the

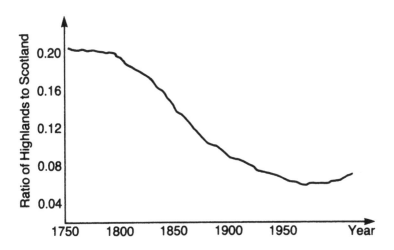

Figure 2.9   *Scottish Highlands related to whole of Scotland.*

Table 2.7   *Population in the Highlands and Scotland 1755–1981*

Year	Highlands (A)	Scotland (B)	$\frac{A}{B} \times 100$
1755	272 225	1 265 380	21.5
1801	324 349	1 608 420	20.2
1811	340 341	1 805 364	18.9
1821	387 605	2 091 521	18.5
1831	414 738	2 364 386	17.5
1841	422 478	2 620 184	16.1
1851	423 880	2 888 742	14.7
1861	407 649	3 062 294	13.3
1871	400 603	3 360 000	11.9
1881	395 980	3 735 573	10.6
1891	389 456	4 025 600	9.7
1901	381 909	4 462 103	8.6
1911	370 220	4 760 900	7.8
1921	365 470	4 882 497	7.5
1931	321 112	4 832 000	6.6
1941		no data available	
1951	314 392	5 096 400	6.2
1961	302 506	5 179 300	5.8
1966	299 789	5 190 800	5.9
1971	306 432	5 229 000	5.9
1974	312 456	5 226 400	6.0
1981	322 594	5 035 315	6.4

individual forecasts equals the total forecast. This is especially so if a variable ratio is being used in the forecasts.

This may seem a more acceptable method than using time as the independent variable as in the trend-line techniques, but the problem arises of how the forecasts for the pattern area are undertaken. These may be trend-line analyses themselves. Nonetheless this is a simple and widely used technique which can be based on elaborate, larger-scale forecasts of the pattern area. The availability of such forecasts is one of the major limitations of the method. Furthermore, the imprecisions and faults of the pattern area forecast are incorporated into the study area forecast. A further problem arises from forecasting the ratio. The techniques employed may, as has already been suggested, be overly simplistic.

A variant of this technique involves an inversion. In such a case the pattern area is a smaller area which forms part of the study area. This method would be used when detailed up-to-date information exists for the smaller pattern area. A forecast for the pattern area might employ a more elaborate technique such as *cohort survival* (see Section 2.3.4).

(b)   Component ratio method

Component ratio methods assume a relationship between a component of the total population and the total population. A component which has been used widely in America is school enrolment. This can be disaggregated by age group. Similarly, number in employment, disaggregated by occupation, may be used. More commonly, births and deaths are used as components (see Section 2.3.3 for a fuller discussion). Each version requires that a ratio or ratios is calculated between the component and the total population, usually from past data. As before this ratio may be constant or variable. From *current* component information, current population may be calculated. It is unusual for this method to be used for forecasts.

(c)   Symptomatic ratio method

Symptomatic ratio methods use, as their base, *objects* or *services* associated with population. Examples are houses, telephones, electricity or water subscribers and vehicle registrations. A constant or variable ratio is used as above. Few of these versions are widely used and most are American in origin. The *housing unit method* is the most popular and useful. The advantage is that information on housebuilding and demolition can usually be readily collected.

Table 2.8   *The housing unit method*

	Owner- occupied	Local authority	Privately rented	Total
Current stock (1981)	100 000	80 000	20 000	200 000
Estimated change (1981–91)	+10 000	−1 000	−5 000	+4 000
Estimated stock (1991)	110 000	79 000	15 000	204 000
Estimated ratio (from past data)	2.0	2.5	2.75	
Forecast population (1991)	220 000	197 500	41 250	458 750

The simplest version of the housing unit method involves estimating the number of housing units and multiplying by the average household size. In Table 2.8 housing information disaggregated by tenure is used to forecast population in 1991. Further disaggregation could be carried out according to size and age of house. The problem of this method is that housing information may be unreliable and it is then multiplied by a further 'estimate' — the ratio. It is possible that this will magnify the separate errors. Such a problem is common to symptomatic methods. The advantage of this method is that population numbers may be calculated for small areas.

Further elaborations of the housing unit method involve disaggregation by type of household, for example:

1   Founding — young household into which children will be born
2   Stable — a family with children at school or no family
3   Contracting — children leaving home to form separate households.

(This version is based on 'typical' family households which, while still dominant, constitute a declining proportion of total households.)

A variant on this (Post, 1969) was used in Manhattan where most households were in category 3. Population was forecasted using the formula:

$$P_{(1958)} = 0.9P_{(1950)} + 3.75 \; \Delta H \{^{(1958)}_{(1950)}\} \tag{2.13}$$

where $P_{(1950)}$ was population in 1950
$P_{(1958)}$ was population in 1958
$\Delta H \{^{(1958)}_{(1950)}\}$ was the change in housing units between 1950 and 1958.

Here it can be seen that with contracting households, population declined by 10% ($[1 - 0.9] \times 100$) over the eight years and an adjustment was made for change in the housing units.

As with trend-line methods the ratio methods described above use a relationship between the independent and dependent variables which may not be *causal*. Population changes are influenced by numerous factors which are ignored. Nonetheless, such methods do force more thought about causal links and may produce better understanding of population change.

2.2.4 Multiple regression method
Multiple regression analysis is based on the assumption of a stable relationship between population and other variables (see Appendix A.9). The simplest form uses only one independent variable as follows:

$$P_{(t)} = a + bX_{(t)} \tag{2.14}$$

where $X_{(t)}$ is the independent variable. It should be clear that the equations for trend-line and ratio methods are variants of this general equation. Once determined this equation may be used for forecasting.

More commonly the equation takes the form

$$\Delta P_t^{t+n} = a + b \; \Delta X_t^{t+n} \tag{2.15}$$

where ΔP_t^{t+n} is the change in population between t and $t + n$.
ΔX_t^{t+n} is the change in the independent variable between t and $t + n$.

A further variant incorporates a *lagging*, that is, the change in the independent variable takes time to have an effect, so

$$\Delta P_t^{t+n} = a + b \ \Delta X_s^{s+m} \tag{2.16}$$

where s and $s + m$ are two dates, chosen as necessary and appropriate.

It is also possible that the relationships are non-linear and so a log transform is necessary before a regression line may be calculated (resorting here to the notation of (2.14) for simplicity). One of the following may be appropriate:

$$\log P_{(t)} = a + b \ X_{(t)} \tag{2.17}$$
$$\text{or } P_{(t)} = a + b \ \log X_{(t)} \tag{2.18}$$

More usually multiple regression is used:

$$P_{(t)} = a + b_1 X_1 + b_2 X_2 + \cdots + b_n X_n \tag{2.19}$$

which may be written in summary form as:

$$P_{(t)} = a + \sum_i b_i X_i \tag{2.20}$$

The variables used may be the actual variables (X_i), the changes in the variables (ΔX_i) with or without a lagging, or a log transform ($\log X_i$ or $\log \Delta X_i$ or $\Delta \log X_i$). Examples of variables are accessibility, cost of housing, wage levels, employment vacancies, unemployment, and industrial invest-ment. In some studies, rather than fit a regression line through historical data for one area, it is fitted through current or recent data for several areas.

The level of complexity of these models masks an underlying theoretical inadequacy. The *causal structure* is poor and the independent variables may conceal numerous specific causal factors linked in different ways to the dependent variable. It also seems unlikely that the parameters (a, b_i) will remain constant — as is required for forecasting. There are also problems which arise from the statistical requirements for using the technique — the independent variables must be normally distributed and independent (see Appendix A.9).

In the past such models were widely used sometimes to predict birth and death rates or components of the population (see Section 2.3.3). More recently they have been out of favour because of the problems identified above. In most cases their accuracy did not justify their complexity. Fore-casting numerous independent variables has many difficulties.

2.3 Composite approaches

The approaches discussed in Section 2.2 are typically used to forecast aggregate population (although ratio and regression methods have been used to forecast components of the population). More usually, and particularly for planning, *compositional* information is required for the population. Details of age, sex and occupational class are often essential in estimating the demand for many services. For this a *composite* approach is required.

2.3.1 Population composition

In any area population change may be accounted for by four components, namely births, deaths, in-migration and out-migration (the last two are often combined as *net* migration). Composite approaches estimate these separately, thus:

$$P_{(t+n)} = P_{(t)} + B_t^{t+n} - D_t^{t+n} + \text{in } M_t^{t+n} - \text{out } M_t^{t+n} \tag{2.21}$$

where $P_{(t+n)}$ is population at time $t + n$
$P_{(t)}$ is population at time t
B_t^{t+n}, D_t^{t+n}, in M_t^{t+n}, out M_t^{t+n} are respectively births, deaths, in-migration and out-migration between times t and $t + n$.

For simplicity assume there is no migration — this will be discussed in Section 2.3.5. If population figures for several past dates and the number of births and deaths during a specified period are known, then population may be forecast using the *crude birth rate* and *crude death rate* (crude refers to the fact that these are for the *total* population rather than by age group).

The *crude birth rate (b)* is defined as the number of births in a time period divided by the average of the population at the beginning and end of the period. The *crude death rate (d)* is similarly defined. The crude survival rate is $(1-d)$, where d is the death rate. Population may be forecasted by applying these rates to the current population, thus

$$P_{(t+n)} = P_{(t)} + b_n P_{(t)} - d_n P_{(t)} \tag{2.22}$$

where $P_{(t)}$ is the population at time t
$P_{(t+n)}$ is the population at time $t + n$
b_n is the crude birth rate for n years
d_n is the crude death rate for n years.

If the population in 1981 is 100 000, the ten-year birth rate is 0.20 and the ten year death rate is 0.16, then the population in 1991 may be forecasted:

$$\begin{aligned} P_{(1991)} &= P_{(1981)} + b_{10} P_{(1981)} - d_{10} P_{(1981)} \\ &= 100\ 000 + (0.20 \times 100\ 000) - (0.16 \times 100\ 000) \\ &= 104\ 000 \end{aligned} \tag{2.23}$$

A variant of this method may be used to estimate current population. For if

$$b_n = \frac{B_n}{P_{(t)}}$$

then $P_{(t)} = \dfrac{B_n}{b_n}$

where $P_{(t)}, b_n$ are as before

B_n is the number of births in n years.

If there are records of numbers of births in n years and the birth rate has been previously calculated, or a national figure is used, an estimate may be made of total population. A similar method may be applied using death rates. As the method ignores migration and assumes constant *vital* (birth and death) rates it is unlikely that the two versions will produce the same answer. In this case an average may be taken. For example if

$$b_n = 0.20,\ d_n = 0.16,\ B_n = 19\ 800,\ D_n = 16\ 200$$

then $P_{(t)} = \dfrac{19\ 000}{0.20}$ and $P_{(t)} = \dfrac{16\ 200}{0.16}$

$\qquad\qquad = 99\ 000 \qquad\qquad\qquad = 101\ 250$

and the estimated population is $\dfrac{101\ 250\ +\ 99\ 000}{2} = 100\ 125.$

This is a variant of the component ratio method discussed in Section 2.2.3.

2.3.2 Cohort survival

The use of crude vital rates can result in substantial inaccuracies as both birth and death rates are dependent on the age structure of the population. Child-bearing is almost exclusive to women in the 15–44 age group, while death rates increase dramatically in the 65+ age group. A population with a large percentage of women of child-bearing age is likely to have a high birth rate, while one with a large percentage of the elderly is likely to have a high death rate. Thus if national crude vital rates are applied to a population with a different age structure, inaccuracies will result.

The *cohort survival* method takes account of the differences in vital rates by using age- and sex-specific rates. A *cohort* in this context is merely an age group. This method produces information on the future age and sex distribution of population. It is the most commonly used technique, particularly at the local authority level. Table 2.9 and Fig. 2.10 (overleaf) illustrate the application of the cohort survival technique to a simple example. The age and sex structure of population shown in Fig. 2.10 is known as a *population pyramid*. The procedure is as follows overleaf:

Table 2.9   *The cohort survival technique*

Male	No.	Survival rate	Survivors			
0–14	20 000	0.85	15 400			
15–29	10 000	0.75	17 000			
30–44	12 000	0.65	7 500			
45–59	5 000	0.50	7 800			
60+	3 000	0.20	3 100			

Female	No.	Survival rate	Survivors	Birth rate	Births
0–14	20 000	0.90	15 400	0	20 000 + 10 800
15–29	10 000	0.80	18 000	2.0	
30–44	12 000	0.70	8 000	0.9	
45–59	5 000	0.60	8 400	0	
60+	3 000	0.40	4 200	0	

(a)   Obtain information on the age and sex structure of the base year population[1] (in this case 1981).

(b)   Obtain/calculate[2] the *survival* and *birth*[3] rates for each cohort.

(c)   Apply the survival and birth rates to the base year population structure.[4]

(d)   Divide births into males and females by applying a *sex ratio* to the total number of births.[5]

Notes

1   In the example shown in Table 2.9 fifteen-year cohorts have been used and a fifteen-year forecast is obtained. More usually five-year cohorts are used and forecasts may be for multiples of five years. If necessary the procedure is repeated (see Section 2.3.3).

2   The rates may be assumed constant or may have to be forecasted (see 2.3.4).

3   As births to women under 15 and over 44 are rare these are typically included in the nearest cohort for calculation of birth rates. Rates outside the 15–44 age group are usually assumed to be zero. Age specific birth rates are sometimes called *fertility rates*.

4   A cohort 'provides' the next cohort for the forecast. For example in Table 2.9 the 15–29 cohort in 1981 becomes the 30–44 cohort in 1996.

5   The births supply the new cohort 0–14 in 1996.

Readers should now apply the procedure to the data presented in Tables 2.10 and 2.11 (overleaf) to produce the forecast for 1991 and use this

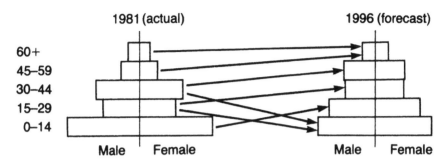

Figure 2.10 *The population pyramid.*

forecast to produce a forecast for 1996.

It is possible to use the cohort survival technique to data disaggregated by occupational class, race, religion, etc. and to use different *vital* rates for each group.

Table 2.10 *Age structure (1986)*

	Total	*Male*	*Female*
0–4	4 100	2 100	2 000
5–9	3 900	2 000	1 900
10–14	4 300	2 150	2 150
15–19	4 000	2 000	2 000
20–24	3 500	1 750	1 750
25–29	3 000	1 500	1 500
30–34	3 100	1 600	1 500
35–39	2 800	1 400	1 400
40–44	2 900	1 400	1 500
45–49	3 100	1 500	1 600
50–54	3 200	1 500	1 700
55–59	2 800	1 300	1 500
60–64	2 800	1 300	1 500
65–69	2 700	1 200	1 500
70–74	2 400	900	1 500
75–79	1 700	600	1 100
80–84	1 100	300	800
85+	600	150	450
Total	52 000	24 650	27 350

Table 2.11   *Birth and death rates*

(a)   Age-specific death rates (for five-year period)

| Age group | Deaths per 1000 males or females | |
	Male	Female
0–4	3.0	2.0
5–9	1.5	1.0
10–14	1.5	1.0
15–19	5.0	1.5
20–24	5.0	1.5
25–29	6.0	3.0
30–34	6.0	3.0
35–39	13.5	7.5
40–44	13.5	7.5
45–49	40.0	25.0
50–54	40.0	25.0
55–59	100.0	60.0
60–64	100.0	60.0
65–69	200.0	150.0
70–74	200.0	150.0
75–79	400.0	300.0
80–84	400.0	300.0
85+	800.0	700.0

(b)   Age-specific birth rates (for five-year period)

Age group	Live births per 1000 females
15–19	150.0
20–24	500.0
25–29	650.0
30–34	330.0
35–39	100.0
40–44	15.0

N.B.   Male births per 1000 female births = 1050

2.3.3 The cohort survival matrix method
Readers who are unsure of basic matrix algebra should read Appendix A.12 before proceeding further.

The calculations undertaken in the cohort survival matrix method can be summarized using matrices and vectors. For simplicity let the population be divided by age into n cohorts, but not divided by sex. Further, let the period of the forecast equal the size of the cohort age interval, that is, if the population is divided into five-year age groups, then the forecast period is also five years.

Let $P_{i(t)}$ represent the population in cohort i at time t, the base year, and let $P_{i(t + 1)}$ represent the forecast population in cohort i. Let S_i be the age specific survival rate, and b_i the age specific birth rate, then

$$P_{i(t + 1)} = S_{i-1} P_{i - 1(t)} \qquad = 2, \ldots, n - 1$$
and $\hspace{11cm}$ (2.24)
$$P_{n(t + 1)} = S_{n-1} P_{n - 1(t)} + S_n P_{n(t)}$$

$P_{1(t + 1)}$ is obtained from births. If P_k and P_l are the child-bearing cohorts, then

$$P_{1(t + 1)} = b_k P_{k(t)} + b_l P_{l(t)} \qquad\qquad\qquad (2.25)$$

(for simplicity assuming there to be only two child-bearing cohorts).

This algebra may be summarized in matrix form. The cohorts at time t and $t + 1$ (the forecast year) may be represented respectively as $n \times 1$ column vectors denoted by

$$\mathbf{P}_{(t)} = \begin{bmatrix} P_{1(t)} \\ \vdots \\ P_{n(t)} \end{bmatrix} \quad \text{and } \mathbf{P}_{(t + 1)} \begin{bmatrix} P_{1(t + 1)} \\ \vdots \\ P_{n(t + 1)} \end{bmatrix}$$

If $\mathbf{S}$ is the *survival matrix* and $\mathbf{B}$ is the *birth matrix*, then

$$\mathbf{P}_{(t + 1)} = \mathbf{S}\mathbf{P}_{(t)} + \mathbf{B}\mathbf{P}_{(t)} \qquad\qquad\qquad (2.26)$$
$$= (\mathbf{S} + \mathbf{B}) \, \mathbf{P}_{(t)} \qquad\qquad\qquad (2.27)$$

where
$$\mathbf{S} = \begin{bmatrix} 0 & 0 & 0 & \cdots & 0 & 0 \\ S_1 & 0 & 0 & \cdots & 0 & 0 \\ 0 & S_2 & 0 & \cdots & 0 & 0 \\ 0 & 0 & S_3 & \cdots & 0 & 0 \\ \cdot & \cdot & \cdot & \cdots & \cdot & \cdot \\ \cdot & \cdot & \cdot & \cdots & \cdot & \cdot \\ \cdot & \cdot & \cdot & \cdots & \cdot & \cdot \\ 0 & 0 & 0 & \cdots & S_{n-1} & S_n \end{bmatrix}$$

and

$$\mathbf{B} = \begin{bmatrix} 0 & \cdots & 0 & b_k & b_l & 0 & \cdots & 0 \\ 0 & \cdots & 0 & 0 & 0 & 0 & \cdots & 0 \\ 0 & \cdots & 0 & 0 & 0 & 0 & \cdots & 0 \\ \cdot & \cdots & \cdot & \cdot & \cdot & \cdot & \cdots & \cdot \\ \cdot & \cdots & \cdot & \cdot & \cdot & \cdot & \cdots & \cdot \\ 0 & \cdots & 0 & 0 & 0 & 0 & \cdots & 0 \end{bmatrix}$$

(Readers should multiply out the matrices and vectors using equation (2.26) and ensure that they can obtain equations (2.24) and (2.25).)

To the non-mathematical the above may seem like obfuscation. In reality it is very simple as can be illustrated by an example. Consider the following:

$$\mathbf{P}_{(t)} = \begin{bmatrix} 40\,000 \\ 20\,000 \\ 24\,000 \\ 10\,000 \\ 6\,000 \end{bmatrix} \quad \mathbf{S} = \begin{bmatrix} 0 & 0 & 0 & 0 & 0 \\ 0.9 & 0 & 0 & 0 & 0 \\ 0 & 0.8 & 0 & 0 & 0 \\ 0 & 0 & 0.7 & 0 & 0 \\ 0 & 0 & 0 & 0.5 & 0.2 \end{bmatrix} \quad \mathbf{B} = \begin{bmatrix} 0 & 1.0 & 0.5 & 0 & 0 \\ 0 & 0 & 0 & 0 & 0 \\ 0 & 0 & 0 & 0 & 0 \\ 0 & 0 & 0 & 0 & 0 \\ 0 & 0 & 0 & 0 & 0 \end{bmatrix}$$

then

$$\mathbf{P}_{(t+1)} = (\mathbf{S} + \mathbf{B})\,\mathbf{P}_{(t)} \tag{2.28}$$

$$= \left(\begin{bmatrix} 0 & 0 & 0 & 0 & 0 \\ 0.9 & 0 & 0 & 0 & 0 \\ 0 & 0.8 & 0 & 0 & 0 \\ 0 & 0 & 0.7 & 0 & 0 \\ 0 & 0 & 0 & 0.5 & 0.2 \end{bmatrix} + \begin{bmatrix} 0 & 1.0 & 0.5 & 0 & 0 \\ 0 & 0 & 0 & 0 & 0 \\ 0 & 0 & 0 & 0 & 0 \\ 0 & 0 & 0 & 0 & 0 \\ 0 & 0 & 0 & 0 & 0 \end{bmatrix} \right) \begin{bmatrix} 40\,000 \\ 20\,000 \\ 24\,000 \\ 10\,000 \\ 6\,000 \end{bmatrix} \tag{2.29}$$

$$= \begin{bmatrix} 0 & 1.0 & 0.5 & 0 & 0 \\ 0.9 & 0 & 0 & 0 & 0 \\ 0 & 0.8 & 0 & 0 & 0 \\ 0 & 0 & 0.7 & 0 & 0 \\ 0 & 0 & 0 & 0.5 & 0.2 \end{bmatrix} \begin{bmatrix} 40\,000 \\ 20\,000 \\ 24\,000 \\ 10\,000 \\ 6\,000 \end{bmatrix} = \begin{bmatrix} 32\,000 \\ 36\,000 \\ 16\,000 \\ 16\,800 \\ 8\,000 \end{bmatrix} \tag{2.30}$$

As in (2.30) the matrices $\mathbf{S}$ and $\mathbf{B}$ are usually combined to produce $\mathbf{C}$ — the *change matrix*. To predict ahead for a further period it is only necessary to multiply $\mathbf{P}_{(t+1)}$ by $\mathbf{C}$, thus

$$\begin{aligned} \mathbf{P}_{(t+2)} &= \mathbf{C}\,\mathbf{P}_{(t+1)} & (2.31) \\ &= \mathbf{C}\,(\mathbf{C}\,\mathbf{P}_{(t)}) & (2.32) \\ &= \mathbf{C}^2\,\mathbf{P}_{(t)} & (2.33) \end{aligned}$$

In the example shown opposite,

$$
C^2 = \begin{bmatrix} 0 & 1.0 & 0.5 & 0 & 0 \\ 0.9 & 0 & 0 & 0 & 0 \\ 0 & 0.8 & 0 & 0 & 0 \\ 0 & 0 & 0.7 & 0 & 0 \\ 0 & 0 & 0 & 0.5 & 0.2 \end{bmatrix} \begin{bmatrix} 0 & 1.0 & 0.5 & 0 & 0 \\ 0.9 & 0 & 0 & 0 & 0 \\ 0 & 0.8 & 0 & 0 & 0 \\ 0 & 0 & 0.7 & 0 & 0 \\ 0 & 0 & 0 & 0.5 & 0.2 \end{bmatrix}
$$

$$
= \begin{bmatrix} 0.9 & 0.4 & 0 & 0 & 0 \\ 0 & 0.9 & 0.45 & 0 & 0 \\ 0.72 & 0 & 0 & 0 & 0 \\ 0 & 0.56 & 0 & 0 & 0 \\ 0 & 0 & 0.35 & 0.1 & 0.04 \end{bmatrix} \qquad (2.34)
$$

The method outlined above may be applied separately for males and females. Readers should now attempt this for the data in Table 2.12.

This basic technique may be adapted to accommodate migration to one area and inter-regional flows of population (Masser, 1972).

Table 2.12   *Cohort survival*

Male	No.	Survival rate (15 years)	
0–14	15 000	0.85	
15–29	20 000	0.75	
30–44	17 500	0.65	
45–59	10 000	0.50	
60+	8 000	0.20	
Female	No.	Survival rate (15 years)	Birth rate (15 years)
0–14	15 000	0.90	0
15–29	22 000	0.80	2.0
30–44	20 000	0.70	0.9
45–59	12 000	0.60	0
60+	10 000	0.40	0

2.3.4 Vital rates — further considerations

It should be clear from the above discussions that the most important part of the cohort survival technique is the assessment of the vital rates. Typically these are calculated from past data for an area, or national rates are used.

These are assumed constant for the period of the forecast. This assumption, although simple, often produces reasonable short-term forecasts but is less acceptable for longer terms as the vital rates can be very unstable.

In longer-term forecasts it becomes necessary to forecast the vital rates. This may be done by using trend lines (see Section 2.2.1), multiple regression (see Section 2.2.4) and by interpretation (often subjective) of a variety of other factors. The forecasting of these vital rates can be problematic. The actual population forecasts depend on the accuracy of the vital rates forecasts and no amount of disaggregation on the basis of occupation, race, religion, income, etc. can compensate for the limitations of vital rates forecasts. The factors affecting death and birth rates are now discussed to show the complexity of the influences on them (for a fuller discussion see Williams, 1977).

(a)   Death rates

The crude death rate has been declining slowly and predictably. Death rates for children under one year are known as *infant mortality rates*. These have been falling and social class variations have been lessening. The death rate for those in the 1–44 age group is less than one in a thousand per year and has been relatively constant. However, because the rate is so low, changes in it make little absolute difference to population numbers. For those in the 45 + age group there are substantial sex differences in the rates. Women outlive men and the male rate in 1971 was approximately the same as the female rate in 1911. However, the increase in the number of the elderly has been more to do with increases in births and decreases in infant mortality rates rather than decreases in the death rate among the elderly. Crude death rates over time are shown in Table 2.13. These rates change very slowly and are accordingly the easiest of the components of population change to fore-

Table 2.13   *Death rates (Scotland)*

Year	Deaths per 1000 population
1973	12.4
1974	12.4
1975	12.1
1976	12.5
1977	12.0
1978	12.6
1979	12.7
1980	12.3
1981	12.3
1982	12.6
1983	12.3

cast for any scale or time period. In most cases they may be assumed constant and errors are small. The main reason for the predictability is that the influences on death rates change slowly and so take a long time to have an effect. Influences include diet, occupation, genetics, and health care.

Age is clearly the dominant factor in explaining differences in death rates. Sex is also important but to a much lesser degree. In comparison to these two, other factors are not of great importance and accordingly national and regional rates may be applied for local area forecasts unless there are any obvious differences. Factors other than age and sex which are of relevance include marital status, social class, occupation, degree of urbanization, and race. These are not all independent factors. In retirement areas with high concentrations of high income groups, or in inner cities with high concentrations of the disadvantaged, variations from national and regional rates would be expected.

Table 2.14 *Birth rates (Scotland)*

Year	Live births per 1000 population
1973	14.3
1974	13.4
1975	13.1
1976	12.5
1977	12.0
1978	12.4
1979	13.2
1980	13.4
1981	13.3
1982	12.8
1983	12.6

(b) Birth rates
Over the last 50 years birth rates have fluctuated enormously (see Table 2.14). Consequently they have been the most problematic aspect of population forecasting, at least at the national level — for smaller areas migration often proves to be more problematic.

The birth rate is the outcome of a set of complex and interrelated factors whose strengths and relative influence on the birth rate vary over time in a complicated manner. These include: proportion of women marrying; age at marriage; timing of births; completed family size; duration of marriage; contraception and abortion. These are now discussed individually.

Proportion of women marrying: The increase in this proportion has been a strong long-term trend. 93% of women are now married by the time they reach 45. A factor in this increase has been the increase in the proportion of men of marriageable age. With other factors constant this leads to an increase in the birth rate.

Age at marriage: There has been a tendency for age at marriage to fall, although there has recently been a reversal in the trend. With other factors constant, a fall in the age at marriage leads to an increase in the birth rate.

Timing of births: The time period between marriage and first birth and the intervals between any subsequent children influence the birth rate. The trend has been for the period between marriage and first birth to increase, but for there to be fewer births later in marriage, so that births are concentrated into a shorter period. The effects of timing of births may be seen in the period after the Second World War. One cohort of women delayed birth mainly because men were in the armed forces. This led to an increase in the birth rate in the years immediately after the war. The effect was coupled with a reduction, in the next cohort, of the period between marriage and first birth.

Completed family size: This is the number of children a couple have during the period of a woman's fertility. It is *not* the same as average family size which includes 'incomplete' families. This is the most important influence on *long-term* trends in birth rates. The average completed family size has remained reasonably constant and only timing has changed. The average for women married only once between the ages of 20 and 24 has been particularly stable. The distribution has, however, become more concentrated with fewer small and fewer large families (although there are most probably different factors influencing these two trends).

Duration of marriage: The numbers of divorces and consequent remarriages have been increasing but the effects on the birth-rate are complex. For example, a remarried couple with custody of both sets of children may decide to have no further children, whereas a remarried couple without custody may have an 'average' number of children in addition to those from their previous marriage. The effect depends on age at divorce, age at remarriage, size of family from first marriage, custody of children, etc.

There has also been a partly deliberate increase in one-parent families and in relationships which constitute marriage in anything other than law. Both are probably concentrated in large cities and the latter is difficult to monitor and so include in analyses. Such trends suggest that completed family size should be superseded by a *generation fertility rate* defined as the total number of children born to one woman.

Contraception: This has become more widely available and socially acceptable. Its effect has most likely been to make family planning decisions such as timing of births and completed family size easier rather than to effect a significant reduction in average completed family size.

Abortion: This has been legal since the 1960s and so widely available, although there are significant local differences in availability. The effect has been twofold: firstly in controlling birth rates in younger women, mainly teenage and unmarried; and secondly in controlling family size.

Each of the above is a factor in birth rates but the modelling of the inter-relationships is particularly difficult. The data requirements for analyses are enormous. For example, annual data on age at marriage, births, etc. are necessary. Furthermore data on *generation fertility rates* are, by definition, dated. Analysis is possible with an assumed constant generation fertility rate and a known distribution of sizes. This leads to an examination of timing of births. The complexity does not end there, for each factor is influenced by numerous socio-economic conditions such as race, religion, social class, level of education, occupation, changing sex roles, cost of living, level of unemployment, cost and availability of housing. Disentangling the labyrinth of complex and interrelated causal relationships is not a task for planners. It is best left to demographers who can provide national and regional level forecasts which may subsequently be used in local area forecasts. For, except at a small area level, the variations are much less pronounced than through time. An awareness of the factors does, nonetheless, assist understanding and allow for local adjustments to be made.

2.3.5 Migration
Until now migration has been ignored in the analysis. Its complexity and importance at the local level require that it is treated separately and in detail. Adjustments may be made to the cohort survival method as necessary to account for migration. This is typically done by simply adding or subtracting the migration estimates from the forecasted figures. Aggregate approaches subsume migration into overall change and do not usually consider it separately.

Of the components of population change, migration, both in and out, is the most difficult to forecast. At the national level in-migration is controlled and so easy to forecast. Out-migration, although not usually controlled by source nations, is controlled by recipient nations and is also relatively easy to forecast. In any case, at the national level, migration is numerically small when compared to births and deaths. By contrast, intranational migration is difficult to forecast, particularly at the local authority level, for spatial variations are enormous. There are also fundamental definitional and data problems which are now discussed.

(a)   Defining migration

It is easy to define a birth or a death, but defining migration is much more problematic. If migration is a move from one place to another, the first problem that arises is defining a move. There are two issues: firstly, how far does the move have to be; and secondly for how long? Is a move next door migration, or a move to an adjoining street, or from one suburb to another? Is a move for one month migration, or a move for a year? These questions may be answered by reference to the spatial scale of analysis and to the period between analyses, but these present further problems.

The spatial scale: In forecasting population in an area the interest lies in moves *in* and *out* of the area and not moves *within* the area. However, the relative importance of moves in/out and within an area depends on the boundaries of an area. For example, in Fig. 2.11 identical levels of movement produce quite different levels of intra- and inter-area migration, depending on the location of the boundary between areas A and B. In general, concentration of population near boundaries results in greater inter-area migration than if population is centrally located. Furthermore, the larger the spatial unit, the larger is the percentage of intra-area migrants.

The period between surveys: The longer the period between surveys, the greater is the total number of inter-area migrants, but the rate of migration for each time unit tends to fall. This is because of the failure to record in-migrants who have died, out-migrants who have returned, and in-migrants who have moved away.

As a consequence of the above problems caution is required when comparing data for different spatial scales and over different time periods. An understanding of these factors is important for forecasting.

(b)   Migration data

National surveys can provide a sound basis for forecasting regional and local vital rates. This is not so for migration data. Not only is migration difficult to define but there are substantial spatial variations. In the absence of population registers which exist in several European countries, the Census provides information on place of birth, previous and present residence. This allows for origin and destination studies but largely precludes temporal analysis. Detailed studies of migration are rare, and are often for small areas. National Health records of registrations with General Practitioners are a useful information source for migration.

(c)   Influences on migration

There are numerous reasons for migration, including employment opportunities; cost, quality and availability of housing; education; retirement; marriage; and socio-cultural facilities. The importance of these factors

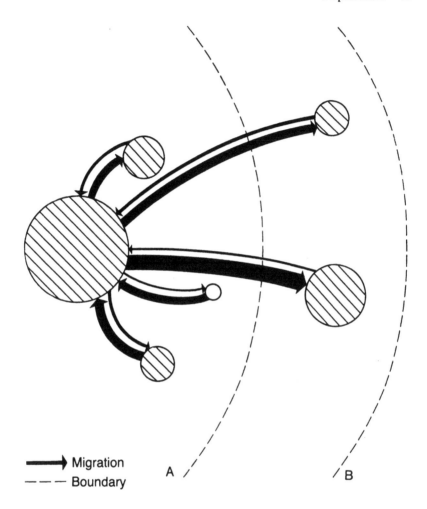

Figure 2.11   *Migration across boundaries.*

varies from area to area and has differential effects according to sex, age, occupation, social class, family structure and tenure of housing.

(d)   Classifying migration

There are several ways of classifying migration. One method is according to the type of area such as urban to rural; suburb to suburb; etc. Another classification which has been used for small rural areas is local in-migrant, other in-migrant, and return-migrant. However, the most useful distinction for local areas, particularly when the problems of defining migration are taken into consideration, is between *short-* and *long-*distance migration.

Short-distance migration   is the more common and can be defined as a change of residence without a breaking of social ties other than perhaps with immediate neighbours. It is predominantly influenced by housing requirements, but may also involve a change in employment. As a household is formed and ages it typically grows in size and then contracts as the children leave home to set up new households. This often coincides with the growth in real income, and so need and ability to pay for different sizes of housing coincide and the household moves. This may involve a change of tenure from, for example, privately rented to owner occupation. Similar movement occurs to and within the public rented sector but movement is more restricted by bureaucratic controls (see Chapter 3).

Long-distance migration   is the less common and involves a break of social ties. It is typically associated with a change in job, although retirement and education are also factors. In this case change in residence is an *effect* of the move rather than the *cause*. Usually the change in job is related to better employment prospects or conditions. It is common in higher social classes and among the young. It is typically from main centre to main centre. Such moves are often easier for those in owner occupation because of the difficulties in transferring council tenancies and finding private rented accommodation. However, tenure of housing is usually related to occupation and long-distance migration is much more common in professional and managerial occupations with 'promotion paths'.

For the reasons outlined above in *(a) Defining migration*, long- and short-distance migration do not necessarily equate to inter- and intraregional migration.

(e)   Forecasting migration

Migration is the most difficult component of population change to forecast at the local authority scale. The results, although often the most important, are least likely to be accurate. The methods used for forecasting migration, whether net, in or out, are similar to those described in the section on *Aggregate approaches* (page 20) to population forecasting.

The first problem which arises is establishing a data base for migration. Even the limited data outlined above is not usually available at the local area level. To overcome this, data on migration is obtained by the *residual method*. This involves forecasting population from vital rates and comparing the results with the actual population figures. Equation (2.22) may be amended to include net migration:

$$P_{(t+n)} = P_{(t)} + b_n P_{(t)} - d_n P_{(t)} + M_n \tag{2.35}$$

where M_n is net migration between t and $t + n$.

Table 2.15 *Estimating migration as the residual*

Male	1971-based forecast for 1981	1981 actual	Net migration estimate (actual − forecast)
0–14	18 000	18 500	+500
15–29	10 000	10 750	+750
30–44	12 000	11 500	−500
45–59	7 500	7 750	+250
60+	5 000	4 500	−500
Female			
0–14	19 000	19 500	+500
15–29	11 000	11 750	+750
30–44	10 000	10 200	+200
45–59	8 000	8 100	+100
60+	7 500	6 750	−750

If $P_{(1961)} = 100\ 000$, $b_{10} = 0.20$, $d_{10} = 0.15$, $P_{(1971)} = 120\ 000$, then
$$M_{10} = 120\ 000, - (100\ 000 + (0.20 \times 100\ 000) - (0.15 \times 100\ 000))$$
$$= 120\ 000 - (100\ 000 + 20\ 000 - 15\ 000)$$
$$= 120\ 000 - 105\ 000$$
$$= 15\ 000$$

In the same way the cohort survival method may be used retrospectively to estimate age specific migration (see Table 2.15). There is, however, an important methodological problem in taking migration as the residual. In calculating birth and death rates no distinction is made between births to, and deaths of, in-migrants, and similarly births to, and deaths of, out-migrants are excluded from the calculation. In effect the births and deaths recorded do not relate to the base population used in calculating rates. In areas where net migration is significant, or it is small but there are large differences between the age structures of in and out migrants, there can be problems (Isard, 1960).

Trend-line methods: A trend line may be fitted to migration data and a forecast made. For migration this method is more limited in usefulness. Whereas births and deaths are relatively stable, in the short term at least, migration is more sensitive and responsive to changes in employment, housing etc. At best, extrapolation should be used only as a check to other methods of forecasting migration.

The comparative method is of limited use in forecasting migration.

Ratio methods: Migration, as with population, can be related, by a constant or variable ratio, to migration to a larger pattern area; to total population change (for example one tenth of increased population is due to migration); to new employment in an area; etc. These methods can be of use when employed in conjunction with other methods.

Multiple regression may be used in the same way as it was used to calculate population change (of which migration is a component) and with the same problems.

Gravity models are based on the assumption that migration varies directly with the size of a force of attraction and inversely with distance, and are discussed in detail in Chapter 5. They require the force of attraction to be relevant, that is to show a connection with migration which may be considered reasonably constant and so able to be used in forecasting. Employment opportunities are often used. Such models are of dubious merit and tend to overestimate short-distance migration — people often commute rather than move home. It should be clear from the earlier discussion of the nature of migration that gravity models misattribute the role of distance. It is likely that there is a threshold beyond which distance is an irrelevant consideration (see Chapter 4).

2.3.6 Conclusion
Migration is not easy to define and data sources are of poor quality. Forecasting is often inaccurate even with accurate historical data, for migration is sensitive to changes in socio-economic conditions and responds quickly to these changes.

A combination of the cohort survival and residual methods is often used in conjunction with trend or ratio forecasts adjusted, often subjectively, to take account of expected socio-economic change. Estimates may be made of the age structure of migrants according to the causes of migration.

2.4 Conclusions

A variety of the more utilitarian methods of population forecasting have been outlined above. Other more complex methods, such as those based on the capacities of an area and the distribution of employment opportunities, have been omitted because they are beyond the scope of an introductory text such as this.

Population forecasting has distinct advantages over other analytical techniques used by planners, principally because accurate historical data is readily available and often in considerable detail. The four components — namely births, deaths, in- and out-migration — are directly measurable and combine in a manner which is comparatively straightforward in theory, although data availability may limit this in practice. This is particularly so when population is divided into resident and migrant categories. Such a distinction is methodologically convenient but may lead to inaccuracies. Understanding of the complexity of the casual factors affecting the components is poor, and furthermore these relationships are likely to change in character over time. Nonetheless the death rate, because it is stable and slow to react to changes in causal variables, is easy to forecast. The birth rate is more volatile, but spatial variations are much less than temporal variation and so national figures can be used as the basis for a local forecast. By contrast migration is the most problematic to forecast at the local scale and local analyses are necessary. Ultimately the 'best' method depends on the purpose of the forecast, the required accuracy, the size of the area, and the period of the forecast. For most local areas a combination of cohort survival using national (or amended) birth and death rate forecasts, and a trend or ratio migration calculation taking into account distinctive local factors, makes an acceptable trade off between complexity and accuracy.

The accuracy of the different techniques was examined by Morrison (1971). He identifies the following problems in making the comparison:

1 The quality of input data varies.
2 It is difficult to compare areas as population sizes and growth rates are themselves variables.
3 It is difficult to devise a measure of 'accuracy'.
4 The periods of the forecasts vary and so the scope for errors from one period to another will vary as external factors change.
5 Techniques are often modified to suit particular local factors.

Nonetheless he concludes:

1 No method shows consistently greater accuracy.
2 Averaging the results of the different methods produces lower errors.
3 Average errors tend to be low for large populations.
4 Average error varies with rate of population growth, being lowest for slow-growing areas, followed by rapidly growing and declining areas.

This suggests that a number of the simpler and cruder methods should be used as checks on the more elaborate techniques. For a planner, in making forecasts, it is necessary to be clear what assumptions are being made and so be aware of what errors may occur through incorrect assumptions or oversimplifications. The robustness of the results should be tested by measuring their sensitivity to changes in the key assumptions and, when necessary, a range of possible outcomes should be produced allowing for divergences. In this way contingencies may be accommodated within planning policies.

3 Housing

3.1 Introduction

Residential land constitutes a greater share of the developed area of cities than does any other single land-use type. Whereas housing comprises approximately 40% of urban land, the next two largest categories, open space and industry, only take up 20% and 10% respectively. Housing is, therefore, an important aspect of land-use planning — and yet planners have only a limited role in influencing housing policy. It is central government which defines standards for dwellings; banks and building societies control private housing finance and determine mortgage rates, and local authority housing departments build public housing, clear slums and rehabilitate where appropriate. Planners tend to influence policy at the strategic level through the development of settlement policies and, more specifically, through the allocation of land for new housing. The allocation of residential land has important ramifications for it is closely linked to employment location, has implications for the provision of transport, and imposes demands on recreational and community facilities. Moreover, for the 'consumer', housing location is important not only in its own right, but also as a means of access to other facilities such as schools and employment.

Structure plan allocations give districts the strategic framework within which to implement policies and these have been a source of disagreement and conflict with private housebuilders. There has recently been much academic debate about the power of planners to influence spatial development patterns and theoretical and empirical work has been undertaken on the influence of landowners and developers. Clearly, central government also has an important role. A discussion of these issues is beyond the scope of this book, but readers are cautioned against supposing that land allocation, either numerically or spatially, is a rational, technical process rather than a complex political one. Nonetheless, allocations are necessary and so, therefore, are forecasts. Before proceeding to examine the techniques used in housing forecasting it is necessary to clarify several terms and to consider policy objectives.

3.2 Definitions and objectives

3.2.1 Definitions

Population: The interest is in the *normal resident population*, that is excluding institutions, the armed forces, holiday home owners, and often students. These groups typically have special requirements and are usually treated separately.

Dwelling: It is not as easy as might first appear to define a dwelling and the Census definition has varied. The definition allows for separateness, independence, privacy, permanence and occupancy. Varying forms of multiple occupancy, therefore, may pose methodological problems. These are often overcome, or overlooked, by using Census data. Typically the term 'house' is used rather than 'dwelling' or 'dwelling house'.

Household: This has been variously used to describe a housekeeping unit, that is a group of people who share household expenses by, for example, cooking together, or the inhabitants of a dwelling. The two definitions are not equivalent. In Britain the former is more commonly used.

Need: Until recently the concept of need has dominated housing provision. In this context need is a normative term, that is, it is determined by minimum standards or norms set (in this case) by central government. (Other types of need are itemized in Section 7.3.) These standards have risen over time as society becomes more affluent. Thus, need (or future need) is the shortfall in quality *and* quantity of the existing (or forecast) housing stock for the existing (or forecast) population, *regardless of ability to pay or preference.* From the outset it is essential to distinguish need from *demand* and from *aspirations.*

Demand: This is usually used to mean *effective* demand, which is an economic concept which reflects the 'amount' of housing for which the population is *able and willing* to pay. (Concepts such as 'latent demand' are discussed in Sections 7.2 and 7.3.) Translating the 'amount' into numbers, type, size etc. of dwellings is problematic (see Section 3.5).

Aspiration: This is different from both need and demand as it reflects the population's 'wishes' regardless of national standards or ability to pay. It has some similarities to *felt need* and *expressed need* as discussed in Section 7.3.

In reality, need, demand and aspirations are interrelated. Aspirations influence both need and demand. Norms are, in part, determined by incomes and prices and these in turn influence demographic factors which affect both need and demand. For the moment the cover-all term 'requirement' is used.

3.2.2 Policy objectives

The objective of housing policy can be put rather vaguely as:

> To provide everyone with a choice in making a decent home at a price they can afford.

When this statement is examined it can be seen that the emphasis is *social* rather than *physical*. It says little about actual physical provision, and most of the terms used require elaboration.

1 What is meant by everyone? Does this mean, for example, all single persons wishing to live alone?
2 How much of a choice is to be allowed in terms of type of house, location, size, etc.?
3 What is meant by decent? What standards are being used?
4 How is the choice to be made? How do allocation procedures operate? Through the market or by waiting lists, etc.?
5 What is an affordable price? What percentage of income should be spent on housing?

Each of these questions poses problems in determining requirements, and the answers are often implicit rather than explicitly stated in the assessments. It is important to be aware of the assumptions being made, as seemingly precise estimates of requirements are typically highly sensitive to changes in these assumptions.

In most versions of housing forecasting there are five stages, each requiring a forecast for a *plan period*, typically five or ten years (see Fig. 3.1).

---

1 Forecast population
2 Forecast numbers of households
3 Forecast the contribution of the existing stock to future requirements
4 Calculate the deficit/surplus from (2)–(3) above
5 Estimate land requirements from (4) above

Figure 3.1 *Forecasting land requirements for housing — a simplified structure.*

Stage one uses the methods described in Chapter 2. It is important to be aware that the *output* of the population forecast, with all its inherent flaws, is typically an *input* into housing forecasts. The other four stages are now examined separately.

3.3 Households

3.3.1 The aggregate method

The aggregate method is the simplest method used to forecast household numbers. The forecast population (P) is divided by the forecast average household size (S) to produce the forecast number of households (H). Thus, when $P = 100\ 000$ and $S = 2.5$,

$$H = P/S \tag{3.1}$$
$$H = 100\ 000/2.5 \tag{3.2}$$
$$H = 40\ 000 \tag{3.3}$$

The population forecast is derived using one of the methods outlined in Chapter 2. Household size may be assumed constant, forecasted using a trend line or similar method, or 'guestimated'.

This is clearly a crude method. Household types and sizes with their differing requirements are ignored. Also ignored are *concealed* households, that is, where a group of people currently form a single household but wish to form more than one: for example, newly weds living with their parents. The advantages of this method are its simplicity and speed of calculation. Its main value is as a check on other more elaborate techniques.

3.3.2 Headship rates

This is, by far, the most widely used technique for forecasting the number of households (see Fig. 3.2). It involves dividing the forecast population into a number of groups, typically according to age, sex and marital status. Further disaggregation according to household size, housing tenure and socio-economic group has also been used. Within each population group a certain proportion will be heads of household. The number of households headed by a particular type of person is obtained by multiplying the number of persons in a group by the proportion who are heads of household (the headship rate). For example, in Table 3.1 (overleaf), there are 4000 males aged 15–39 and married, and the headship rate is 0.75. This gives 3000 households. This calculation is repeated for each group in the disaggregated population. The numbers of heads of household in each group are then summed to give the total number of heads of household and thus the *forecast* number of households.

---

1 Calculate headship rate from base year population
2 Forecast population, disaggregated as appropriate
3 Forecast headship rates
4 Calculate the forecast number of households
5 From the forecast households calculate potential households

---

Figure 3.2 *The headship rates method.*

Table 3.1   *The headship rates method*

Population group		Population	Headship rate	Forecast households
Married				
Males	15–39	4 000	0.750	3 000
	40–59	5 000	0.950	4 750
	60+	2 000	0.975	1 950
Females		11 000	0	0
Single, widowed & divorced				
Both sexes	15–24	4 500	0	0
	25–39	2 000	0.100	200
Widowed & divorced				
Males	40–59	200	0.675	135
	60+	600	0.650	390
Females	40–59	600	0.750	450
	60+	1 800	0.667	1 200
Single				
Males	40–59	500	0.260	130
	60+	250	0.380	95
Females	40–59	1 000	0.290	290
	60+	600	0.450	270
Total		34 050		12 860

Note: The method assumes that the husband of a couple is the head of household, and that those under 15 are dependents.

Future headship rates are obtained from headship rates calculated from Census data for the base year of the forecast. These are then assumed constant or extrapolated by trend line to produce the forecast headship rates.

From the calculation of the forecast number of households, it is necessary to calculate the number of *potential* households. This allows for *concealed* households, that is, those not currently forming a household but wishing to. It also makes allowance for existing separate households with an assumed preference to share. The conventional calculation is:

Potential households = forecast households
                  + number of married couples not heading a
                    household (it is assumed that they wish to)
                  − three quarters of those single person house-
                    holds sharing a dwelling with another
                    household (it is assumed that they have a
                    preference to share)   (3.4)

In effect this turns households into a dwellings requirement (but see Section 3.2.1 on problems of defining a dwelling). It must be emphasized that this is little more than convention, and its basis, either empirically or behaviourally, is tenuous.

An example of a more detailed disaggregation of headship rates is shown in Table 3.2(a). Readers should now apply these headship rates to the population forecast produced in Section 2.3.2. The following stages are involved:

1. Divide the forecast population into age/sex/marital status group by applying the information in Table 3.2(b) (overleaf) to the forecast.
2. Apply the headship rates of Table 3.2(a) to the disaggregated population (obtained from 1 above) to produce the *forecast* households.
3. Calculate *potential* households from equation (3.4) (assume 1000 single person households share a dwelling with another household).

(a) Problems with the headship method

1. Data problems
The input data for the headship method is the population forecasts which suffer from all of the problems outlined in Chapter 2. Errors in death rates can have a significant effect on the numbers of elderly and hence on numbers of elderly households. Birth rate errors, by contrast, influence the *size* of household rather than number of households, as children will not be involved in household formation until they are at least 15.

2. Modelling problems

Headship rates: Headship rates are assumed constant or predictable. Marital status has to be predicted, itself no easy matter. Over recent years headship rates for population groups have been changing, and these changes have a significant influence on the total number of households.

Table 3.2   *More detailed headship rates data*

(a)   Headship rates — 1981

	Male			Female		
Age	*M*	*S*	*W/D*	*M*	*S*	*W/D*
15–19	0.25	0.02	—	—	0.02	—
20–24	0.80	0.09	0.30	—	0.10	0.50
25–29	0.90	0.25	0.50	—	0.30	0.60
30–39	0.95	0.70	0.75	—	0.60	0.75
40–64/59	0.97	0.70	0.75	—	0.70	0.75
65/60–74	0.95	0.70	0.90	—	0.70	0.90
75+	0.65	0.70	0.80	—	0.80	0.85

(b)   Marital status by age cohort — 1981

			% ages			
	Male			Female		
Age	*M*	*S*	*W/D*	*M*	*S*	*W/D*
15–19	5	95	—	5	95	—
20–24	30	70	—	30	70	—
25–29	65	30	5	65	30	5
30–39	75	5	20	75	5	20
40–64/59	75	5	20	72.5	5	22.5
65/60–64	75	5	20	50	5	45
75+	70	5	25	30	5	65

Note: M = married; S = single; W/D = widowed or divorced

Disaggregation: The headship rates method makes the assumption that the categories used (for example, single males aged 15–39) are sufficiently disaggregated to allow use of standard headship rates. In reality, it is very likely that headship rates differ for different sizes and types of household, and from area to area. Further disaggregation poses problems for data availability, and is, in any case, of questionable value (see below).

Furthermore, account is not taken of household types other than the 'traditional' family or singles. Other household types are subsumed in these rates. Yet it is just such households which contain vulnerable groups and are on the increase.

Causal relationships: The method uses the age/sex/marital status of the population as *independent* variables to explain household formation variations, or, at least, assumes that these demographic factors are modified in predictable ways by tradition, attitude, choice, etc. However, household formation depends not only on numbers and type of person in household forming groups, but also on housing availability and complex socio-economic factors. Indeed, it could be argued that household formation rates should be forecast from housing supply, rather than used as an input in estimating housing shortage. A shortage of housing is likely to be a limiting factor in population growth and household formation.

More generally, housing availability, population growth, household formation, employment and other socio-economic factors such as income interact in a complex manner. Yet migration, which is interlinked in a labyrinth of causal relationships, is typically taken as a *residual* and used as an *input* to housing forecasts (see below).

3.3.3 Other methods
A number of other methods have been developed to forecast household numbers, but none has been widely used.

(a)   Index method
This is a variant of the ratio method used in population forecasting. For example, the ratio of population 20+: private households has remained remarkably constant over the last 40 years. If a value of this ratio (R) is assumed or predicted it can be applied to the forecast population (P) to give the forecast number of households (H):

$$H = PR \tag{3.5}$$

If $P = 4000$ and $R = 0.50$, then

$$H = 40\,000 \times 0.50 \tag{3.6}$$
$$H = 20\,000 \tag{3.7}$$

This is a simple and crude method whose limited value is as a check on the headship method. It gives no idea of the underlying causal mechanisms.

(b)   Life tables
The method involves complex modelling of household 'progress' from a base year. Households are disaggregated into, for example, those which are expanding in size, declining, or stable depending on their stage in the 'life cycle'. Newly formed households consisting of a couple would be expected to increase in size as children are born, then to remain stable before decreasing in size as the children left home.

The effort involved in this method and the absence of explicit consideration of the relationships of household formation to housing supply and other
socio-economic factors render the technique of limited value. Its merit is
that it forces examination of the dynamics of household formation and size.

(c)   Regression
This method is discussed in the following section on household size.

3.3.4 Household size
Household size is clearly an important aspect in any forecast of housing
requirements. It is insufficient, though all too common, to calculate only a
total number of households. There are several ways in which household size
may be taken into account:

1   Typically, household size is ignored at the forecasting stage. Private
    builders or the public sector determine the mix of house sizes according
    to a perception of marketability or need based on past experience,
    policy factors, or an assessment of waiting lists. This has proved to be
    unreliable as it has concentrated on the 'traditional' family, thus
    underproviding for single persons and other groups.
2   An improvement is to assume that the proportions of different
    household sizes are constant and so apply base-year figures. This is a
    simple method, which is better than ignoring the issue, but which itself
    ignores the dynamics of household formation and other demographic
    factors.
3   Another method is to use trend lines for each size of household and for
    average household size. A 'best fit' answer is then produced by
    minimizing the sum of squares of differences, subject to the proportions
    in each size summing to unity and the predetermined average
    household size. This method, akin to regression (see Appendix A.8),
    reduces complex relationships to suspect algebra and is, therefore, of
    dubious value.
4   One final method is multiple regression. The proportions of households
    of each size are taken as the dependent variables. The independent
    variables are demographic, such as the male/female ratio;
    children/mothers ratio; socio-economic group; activity rates; etc. The
    result is the proportion of households of different sizes.
        As is always the case with multiple regression, it is assumed that the
    relationships between independent and dependent variables are
    constant over time. The usual problems of multiple regression apply
    (see Section 2.2.4 and Appendix A.9), and its use reflects an earlier
    preoccupation with such modelling techniques — now, fortunately,
    treated more sanguinely.

Following this discussion of the forecasting of numbers of households, the next section examines the calculations involved in estimating the number of houses likely to be available at a future date, and the consequent requirement for new building.

3.4 Houses

3.4.1 Estimating future stock

The next stage of the procedure is to forecast the contribution that the existing housing stock will make to future requirements. This is done by a simple calculation as shown below:

Future stock (S_F) = Existing stock (S_E)

      − existing substandard housing   (a)

      − houses likely to become substandard
        in the plan period   (b)

      + existing substandard housing which
        will be improved in the plan period   (c)

      + houses likely to become substandard,
        but which will be improved in the
        plan period   (d)

      − other losses   (e)

      ± conversions   (f)

      − second homes   (g)

      − vacant houses.   (h)

$$(3.8)$$

A figure for existing stock is usually easily obtained as such information is collected for rating purposes. Other aspects are more problematic as these depend on standards set, policy attitudes and available finance. Many of these issues are outside the control of planners and to a large extent unknown in the future. The individual items in (3.8) are now discussed.

(a)   Substandard houses
In Scotland the term used is 'below the tolerable standard' while in England and Wales it is 'unfit'. The definitions of these terms have varied over time. Information is most easily calculated from a housing condition survey, though this is likely to be restricted to areas where poor quality housing is concentrated rather than to be comprehensive.

(b)   Houses likely to become substandard
This aspect may be calculated by use of a housing stock ageing model which assumes that houses become substandard at a certain age. Another

technique which has been used is multiple regression, with the number of substandard houses as the dependent variable, and the independent variables as some combination of repair costs and housing characteristics. As costs of repair and the ability to pay change, hence the amount of repairs undertaken changes. However, in many cases it is probable that the calculation is not rigorous, and involves 'guesses' and 'local knowledge'.

(c) Existing substandard houses to be improved

(d) Houses likely to become substandard but which will be improved
These categories, more than the others, are subject to resource and other policy decisions. At times the dominant policy has been the demolition of older properties, resulting in a requirement to provide new houses, whereas more recently policy has focussed more on renovation of the existing stock. Clearly the importance of improvement policies depends on the type of area under consideration. It should be noted that category (d) requires consideration of category (b).

Phasing of renovation and demolition policies is of relevance to new housing requirements within the plan period. A 'rolling' long-term programme will have a quite different effect from a 'blitz' on substandard housing.

The extent of the problem will influence policy as demolition requires replacement new building which depends on resource availability. More detailed aspects involve prioritizing by area or type of house which can influence land availability for new building and the location and type of houses required in the plan period. Factors such as existing housing density, house size and characteristics of households are relevant.

Ideally, policy options should be considered and costed, taking into account differing clearance rates, new-building densities, improvements, differing building rates, functional obsolescence (and hence conversions — see (f) below), etc. Such detail is rarely considered and approximate figures are produced instead.

(e) Other losses
This category includes loss of good quality houses because of policy factors such as new road building. In most cases the number of houses involved is small and relatively easy to calculate.

(f) Conversions
Conversions can result in either a loss or a gain to the number of houses. Large houses may be converted to flats, or bed-sits converted to family houses. The importance of conversions and their effect on stock will vary substantially from area to area. It is most likely to be important in inner area or 'twilight' zones.

(g) Second homes
In many rural areas or coastal resorts, and in international centres such as London, second homes can be an important feature. Estimates can be made from surveys or from valuation rolls.

(h) Vacancies
As households move from one house to another it is probable that the vacated house will remain vacant for a short period, even if only for a few days. This means that a percentage of the total stock will always be vacant. This 'natural' vacancy rate is an inevitable and necessary aspect of household mobility. Conventionally, this is taken to be 3%, but detailed surveys have suggested wildly differing figures from area to area and over time. Figures as high as 10% have been calculated. Clearly net housing requirements can be influenced by the figure used in calculation.

The final calculation of future housing stock, being a combination of the above estimates and guestimates, is at best a rough estimate, and at worst an almost arbitrary figure. It should never be taken as an accurate figure and ideally a range of figures should be calculated.

3.4.2 Problems of the 'accounting' technique
Having forecast the number of households and the contribution of existing stock to housing requirements, the new housing requirement is simply the difference between these two figures. In the unlikely event of this figure being negative there is a forecast housing *surplus*. More likely there will be a shortfall.

It should be clear that the method outlined above is a crude accounting framework. It ignores size, tenure, quality, type and location of the houses, and the availability of amenities and ancillary services. Both houses and households are in effect regarded as homogeneous groups. The technique can be elaborated to include household and house size, but this is rarely done in practice and, in any case, still ignores a wide range of other issues listed above. Nonetheless, this crude accounting framework is the basis for most local authority housing forecasting.

(a) Allocation
A more fundamental problem with the technique is its total failure to consider allocation of households to houses. It is assumed that new building and a 'filtering' through the existing stock will produce an equilibrium in which household requirements and houses are in balance. In reality it is likely that new houses will be taken by in-migrants or by those already well housed. The provision of new housing does not imply the satisfaction of either need or demand. Neither market nor local authority allocation procedures are suitable for some groups. In the accounting method, requirements

are seen only in the physical terms of number of houses, and not in social terms. Without detailed and costly surveys of those obtaining new housing, only crude numerical monitoring is possible. It is extremely difficult to see if actual households in need are allocated new housing. Moreover, even if data were available on household size and house size, and if these matched in aggregate numerical terms, a match could not be deduced because of the possibility of overcrowding and underoccupancy.

(b)   Self-fulfilment
A calculation of a housing requirement can become a 'self-fulfilling prophecy'. If a requirement is calculated and the houses built it is most probable that these will be filled from in-migration, moves up market, and an increased rate of household formation.

3.5 Land supply, availability and allocation

3.5.1 Introduction
Having established by the crude accounting procedure described above the number of new houses required, it is then necessary to convert this into allocations of land. This raises a number of issues such as building densities (and hence the area of land), availability of land, location and tenure split. (This final aspect is considered in detail in Section 3.6.) It should be clear that such issues cannot be resolved by the application of a rational technical methodology, but rather require a policy, and thus political, framework.

A crude calculation can be undertaken to establish the area of land required:

$$A = \frac{H}{D} \tag{3.9}$$

where A is the area required
   H is the required number of houses
   D is the building density.

For example if H is 1000 houses and D is 20 houses per acre, then

$$A = \frac{1000}{20} \text{acres}$$
$$= 50 \text{ acres.}$$

Either gross density (the whole site) or net density (excluding roads, open space, etc.) may be used. It is much more likely that densities will vary from site to site and will be established on an individual basis. This will involve negotiation with builders to agree site capacities subject to policy considerations, site characteristics, type and size of houses, surrounding amenity, etc.

Location is an important strategic issue and it is unlikely that planners and builders will coincide in their views on preferred locations. Planners, particularly over recent years, have a preference for redevelopment of inner area sites often with infrastructural and environmental problems. Builders, in most circumstances, prefer suburban greenfield sites which are, or are perceived to be, more 'marketable' and where production costs are lower. They are likely to want to maximize density and thus profit within any market band or price range.

Tenure split is also an important policy consideration. Recently, central government funding for new council housing has been reduced and the emphasis has shifted to private housebuilders. This has implications for land allocation, type of housing and density, as it may involve a move 'down market' by private housebuilders so as to provide for lower income groups who are unable to obtain council housing. Some groups will not be able to afford this option. Such issues can only be decided within a strategic planning and financial framework which will depend on wider central and local government policy rather than on planning policies for land allocation. This policy framework varies from period to period and from place to place.

This is not to suggest that the resolution of these policy issues can provide a framework within which a purely technical methodology can be applied to assessment of land availability and allocation to housing. There is no standard technique for assessing land availability and the subject is hindered by definitional and interpretative disagreement both between the different interest groups and also between local authorities. Patterns of ownership, for example, influence allocation and may restrict availability for development. Much of this subject area is beyond the scope of this book, but the material is usefully and comprehensively covered in Hooper (1985).

The procedure outlined below is an amalgam of various techniques and is summarized in Fig. 3.3. Land supply is planned for a five-year period; thus, at any time, as land is developed new land is allocated to ensure a continuing supply. For local authority housing most of the procedures outlined below may be internalized within the authority or are not relevant. Attention is therefore focussed on the private sector. The way in which allocations between the two sectors are determined is discussed in Section 3.6.

1   Establish plan period
2   Establish *local market areas*
3   Calculate *established* land supply
4   Calculate *effective* land supply

Figure 3.3   *Estimating land supply.*

3.5.2 Local market areas

A local market area is one within which houses are 'reasonable' substitutes, or within which households are 'reasonably' indifferent to location. These definitions vary substantially from household to household depending on preferences, socio-economic group, income, mobility, place of employment, etc. The supply of housing also influences household definitions. At one extreme it is clear that London and Glasgow do not form part of a local market area. Whether different commuter settlements within a district council area are or are not part of the same local market area is more open to debate. As the objective is to match supply of and demand for land within each local market area, local authorities are more likely to regard these areas as larger than builders do. This gives more flexibility in allocations and avoids allocating land in areas of high amenity or where there are other development pressures. By contrast, builders are more likely to want to have smaller local market areas as this gives more scope for pressing for allocations in particular areas, or in making the case for specific sites.

3.5.3 Established land supply

The terms *established* and *effective* land supply are suggested by SSDP/COSLA (1984). They are not widely used, but are an attempt to categorize and standardize the methodology used in calculating land supply. The approaches of different local authorities vary considerably, but most consider some or all of the factors discussed below.

The established land supply includes the following categories of sites:

1   Sites currently under construction but with capacity remaining.
2   Sites with planning permission.
3   Adopted local plan sites without planning permission.
4   Sites which have a potential agreed on by developers and the planning authority. (These may include inner area sites or buildings for conversion into housing.)
5   Unallocated sites which become available through plan amendment, development control decisions or on appeal. Some of these can be anticipated and included in calculations of future supply, but most cannot as they will involve policy change, often forced by central government. Such sites may be important factors in new land provision, particularly in areas of development pressure.

Sites in categories 1 and 2 will have their capacities for numbers of houses agreed between the developers and the planning authority. Categories 3 and 4 require either allocation by the planning authority or preferably agreement with the developer, which may prove problematic. Hooper (1985) suggests developers may enter agreements to ensure site designation and may thereafter seek an increased capacity.

Two further types of site may be included. Firstly, *small* sites, usually taken to mean sites with a capacity of under five units. These are difficult to identify and if considered at all a calculation is based on past trends and existing planning permissions. Secondly, *windfall* sites, which are completely unexpected sites. These sites can only be considered on the basis of past trends and can be a controversial category which planning authorities may use to justify a lower land allocation for identified sites.

3.5.4 Effective land supply
The above calculations produce what has been termed the established land supply. From this may be calculated the effective land supply. This takes into account the following factors:

1 *Physical and infrastructural constraints*
   The timing of infrastructural provision by the local authority may affect the availability of all or part of a site within the plan period, even if a site has planning permission. Remedial work to inner area sites with physical constraints may have a similar effect.
2 *Ownership constraints*
   If a designated site is not owned by a developer there may be delays caused, for example, by a disagreement over valuation. It is also possible that larger developers may amass considerable 'land banks' as reserves and so 'squeeze' out smaller builders who are unable to compete for sites and so suffer a land shortage. How, and indeed if, such factors should be taken into account is problematic, particularly given the current emphasis on 'market freedom'.
3 *Marketing constraints*
   A builder may prefer not to develop a large site over a short period to avoid 'flooding' the market and so reducing profitability. Thus, although a site has planning permission, adequate infrastructure, and is owned by the developer, it may not be regarded as completely developable within the plan period. As with 2 above, how and if planning should accommodate such considerations is a matter for debate and no rigorous method can be outlined. Moreover the planning controls over phasing are weak. Phasing affects both the rate and the amount of development. The outcome is a matter for central government guidance, local authority politics and negotiation with developers.

From the above considerations land supply can be calculated, and further land allocated if the supply is insufficient. It is usual to allow for an oversupply to allow flexibility. It should be clear that the technique is not precise but is subject to methodological problems and to the effect of interest groups with power to influence decisions.

Land allocation can reduce the price of land, which in turn increases demand and so uses the allocation. Release can therefore create a self-fulfilling demand. As before there is a circularity in the causal relationships between supply and demand, and also between population and housing forecasts.

3.6 Need, demand and aspirations

Until now the distinctions between need, demand and aspirations have been sidestepped in the discussion of forecasting methodology. This is because Sections 3.2–3.4 dealing with a demographically based requirement do not require the distinction to be made. Section 3.5 deals with private sector land allocation and thus demand. The distinction between need and demand, in particular, is an economic one and is an important factor in the tenure split in allocations of land for new housing. The differences and ways of forecasting each are now considered.

3.6.1 Need

Once an estimate has been made of overall need, it is necessary to determine how much of this should be met by the public sector. Public sector housing, mainly provided by local councils, may be regarded as being for those unable to obtain adequate housing in the private sector, because the market will not provide it at a price they can afford. Of course the matter is not as simple as this as many households provided for by the public sector could afford to purchase a house. This is partly explained by the fact that the public sector was once seen as an alternative tenure, although in practice it is now reduced to more of a 'residual' status. An element of preference and tradition is therefore included. In the past local authorities have also housed those displaced by redevelopment and the loss of privately rented accommodation.

The minimum standards basis for need clearly influences its level. At one extreme if the minimum standard was a tent, it is probable that most households could afford this. At the other extreme, if it were a mansion house, then few could afford it. Moreover, policies on rent control, improvement grants, mortgage interest tax relief, etc. influence the demand for private sector housing and hence the requirement for council housing. For example, if tax relief were removed from mortgages fewer people could afford the cost of a mortgage and would have to rely on other sectors (unless house prices fell). The availability of different types of housing tenure also influences demand.

In practice public sector housing provision is influenced by political decisions on expenditure and priorities in resource allocation both at central and local government levels. Recently expenditure has been rather tenuously based on the calculation of need. Nonetheless, the following methods have been used to calculate need.

1 *The shortfall calculation*
   One method is to estimate public sector requirement as the expected
   shortfall in private provision. This assumes reasonably accurate
   forecasts of both total need and demand (see Section 3.6.2). It is, at
   best, a useful starting point.

2 *Financial possibility*
   The opposite approach to 1 above is to calculate what it is financially
   possible to build and, in effect, assume that other sectors will supply
   the rest. This is, of course, an absurd basis for public needs forecasts,
   as available finance should be related to need. Nonetheless, with
   recent cutbacks in public sector housing finance, public provision has
   tended to follow this approach, particularly as the emphasis has been
   on encouraging owner occupation.

3 *The council housing waiting list*
   This often provides the substantive information in the assessment of
   public sector need. As a data source, a waiting list is flawed in a
   number of ways:
   (a) it includes many who are able to afford owner occupation, but
        prefer not to;
   (b) it includes those merely seeking a change of house and currently
        'adequately' housed;
   (c) it may give double counts in adjoining areas;
   (d) it is invariably out of date, and includes those already rehoused;
   (e) it reflects availability of council housing — households do not
        bother to register if they think there is little chance of success;
   (f) it reflects council policy in provision and so excludes significant
        groups 'in need'.

4 *Special needs groups*
   Rather than consider aggregate need, some housing studies have
   chosen to concentrate on special needs groups as a priority. Examples
   are large families, the elderly, the single, students, the homeless, the
   handicapped, etc. The groups are first of all identified and then
   quantified. Quantification is often difficult and sometimes *ad hoc*. It
   can be done from waiting lists, household projections, surveys, social
   work data, or forecasts as a percentage of population. It is necessary in
   such surveys to be clear as to the purpose. It is possible to select the
   groups because they cannot afford to obtain housing in the private
   sector or because of their requirement for a special *type* of housing.
   The latter need not necessarily imply an inability to pay. In cases of
   special need it is necessary to consider access and allocation, which
   implies establishing priorities in letting policy.

3.6.2 Demand

Until now the analyses presented have worked 'up' from demographic data to demand. The following section discusses models which work 'down' from economic data to population. Demand implies a willingness and an ability to purchase housing. It should not be confused with need or aspirations. Nonetheless, the term is often used loosely to refer to either of these. Need and demand are interrelated in a complex manner, as need is normatively defined according to assumptions about levels of demand. Standards or norms for need would not be set at levels which did not reflect standards which were affordable by both households and government. Both are linked to aspirations.

(a)   Why consider demand?
At the most basic level it is necessary to examine demand as, once having established need, the difference between it and demand can be the basis for public sector provision. Until recently this has been true at both the national and the local level. At the national level housing demand forecasts are an input into macro-economic models of the economy. At the local level demand assessment should provide the basis for land allocation to private sector housebuilders.

(b)   The housing market
Before examining methods of assessing demand, it is necessary to outline certain aspects of the housing market. It does not fit well with the ideal 'neo-classical' market for numerous reasons. The good for sale or purchase is not homogeneous. Houses are not alike: they differ in quality, price, size, type, location, tenure and associated facilities, etc. It is, therefore, difficult to establish a unit of housing for measurement of demand, rather than number of houses. Money is often used, but converting this to a number of houses is difficult. Houses are durable, have a high capital cost which usually requires borrowing, and may be rented. They are usually bought by families rather than individuals. As houses have a use and an exchange value there is a very important secondhand market. For these reasons it is necessary to consider submarkets by price, location, etc., rather than conceive of a single market. Further, adjustments to supply take place slowly. A change in prices requires to be high enough and for long enough to encourage new building beyond existing levels. There is a considerable delay between a decision to build and completion. It is also likely that in most areas there is an element of monopoly in that there is a small number of large builders and a limited number of owners of developable land. Such factors profoundly influence location and quantity of new housing. One final, and obvious, feature worthy of mention is the substantial state intervention in setting standards, providing finance and in regulating land use.

(c)   Models of demand
Models of national housing demand have been developed but are of little use to planners in a local area. By contrast local area models are not well

developed and of little practical value in local forecasts. The former are outlined to show the main relationships, while the latter are sketched to show their limitations and how they might be developed.

National models These involve forecasts of economic growth; income distribution (with an assumed percentage spent on housing); housing costs; policies such as tax relief and rent control; etc. In general the approach is to study income elasticity, that is changes in demand resulting from changes in income. For forecasts over longer periods household formation rates are considered.

There are broadly two ways of dealing with the fact that houses are not homogeneous goods.

1 *The housing service approach*
  This assumes that there is a single homogeneous good — housing service — which individual houses yield in differing quantities.
2 *The 'hedonic' approach*
  This assumes that consumers derive utility from the many different attributes which the good — in this case the house — contains rather than from the good in itself. The composite good may be disaggregated into component parts. This involves complex statistics.

There are several problems with this approach. Firstly, it is difficult to define income — it could be gross, net or disposable; household or head of household. Assumptions must also be made about the percentage of income to be spent on housing. Problems also arise because a house has (usually) a rising real value and hence an 'income'. It is an investment and this influences purchase decisions. (This is partly overcome by the 'hedonic' approach.) There are also problems of cross-elasticity of demand between submarkets and between repair and new purchase. Even if an accurate measure of future expenditure on housing can be obtained, the problem remains of turning this into a number of houses. Typically it is assumed that relationships are constant and so demand can be predicted. More fundamentally there is no identifiable *dependent* variable if house sales are used, as these do not equate to demand but are the product of the interaction of supply and demand. As with all such models, forecasting the independent variables poses problems. Models such as those sketched above are of no value at the local level. Quite apart from the modelling problems, the data necessary for their use do not exist. Even if they did, the problems of identifying local market areas and submarkets within these are probably insoluble (see Section 3.5.2 above). Local models are now discussed.

Local models
1 *Past rates*
  One commonly used but decidedly limited technique is to apply trend

lines to past building rates. This ignores all *causal* variables and issues such as migration and supply-led demand. Further, the trend lines vary according to the level of spatial aggregation. A high level of aggregation can easily give over- or underestimates for smaller areas. (This is similar to the problem of housing market areas (see Section 3.5.2).)

2 *Zoned land*
The amount of land zoned for housing by a planning authority has been used as an estimate of future completion rates which are equated with demand. The circularity of argument involved is bemusing. Further, determining densities to convert land to dwellings is not without problems.

3 *Builders' assessment*
It has been argued that as builders are 'close to the market' they will have a reasonable assessment of demand. Such assessments are likely to be 'hunches' rather than forecasts. It should also be noted that private housebuilders are in competition and their aggregate land requirements are likely to be based on over-optimistic assessments of their market shares. Their views cannot be regarded as objective (see Coopers and Lybrand, 1985).

4 *Local studies*
In some areas, studies have been undertaken based on questionnaire surveys, usually of movers. Information is collected on changes in income, motives for moving, household characteristics, etc. Such techniques are costly and time consuming. The causal relationships are not yet adequately developed. Further, it is assumed that the answers to the questions are 'reasonable'; that preferences are not based on misconceptions of, for example, price structures or on limited knowledge of options. The output is a possible, rather than an actual, demand.

None of the above techniques is likely to be particularly accurate as none has a sound theoretical base. A further consideration is aspirations.

3.6.3 Aspirations

Aspirations represent wishes regardless of demand and supply. They are often an implicit aspect of housing provision and interact in a complex way with demand and need. Aspirations do not develop in a vacuum, but are culturally and socially formed. They are worthy of consideration, but difficult to assess. Most housebuilders would claim an awareness of 'what the customer wants'. Aspirations may be assessed in a number of ways (Niner, 1976).

1 *Reasoned intuition*
In effect this assumes a transfer of middle-class values to new home owners. At times it is little more than guesswork.

2 *Questionnaire*
   As in item 4, page 74.
3 *Ad hoc*
   From preferences for, e.g., house type (as expressed in waiting lists) or
   from price preferences in the private sector.
4 *Public pressure*
   This is a somewhat vague approach but is worthy of consideration at
   the local scale if such pressures have been expressed. (This is similar
   to expressed need — see Section 7.3.)

Clearly the above overlap and none is a rigorous method. The methods may
be disaggregated by family type, social class, etc. Within the context of a
comprehensive and coherent policy framework, aspirations are worthy of
assessment. Such a framework rarely exists however — policy is the outcome
of political constraints, interest-group pressure and *ad hoc* decision making.

Some combination of the methods outlined in Sections 3.6.1–3.6.3 is used
to determine the tenure mix in new housing. The shift to owner occupation,
for reasons of aspiration and government policy, leads to the further consideration of *tenure change*. It can be assumed that preferences and demand for
owner occupation among new and existing households will continue to
increase at constant rates and land made available accordingly. Alternately,
changes can be assumed in these rates over time. It is also possible to
disaggregate to, for example, 'housing market areas' and to apply different
rates depending on the existing tenure structure. However, this is an issue of
dispute between housebuilders and planning authorities (see PEIDA,
1985).

3.6.4 Comprehensive analysis

A comprehensive approach to housing policy at the local level rarely exists
in practice and the methodology is not fully developed. Niner (1976) suggests that it might take the following form:

1 *Identify* the actors in the process
   • *demanders:* by income, class, length of residence, family/household
     size, age, type of housing.
   • *suppliers:* local authorities, housing associations, developers,
     landlords, owner occupiers.
2 *Categorize* into broadly similar groups with predictable behaviours.
3 *Identify* demand–supply relationships, goals, activities, capacities for
   acting in the market, allocation policies, finance, etc.
4 *Establish* hypotheses on current and future behaviours.
5 *Identify* data sources to test hypotheses. To overcome surveys such an
   approach requires a parallel improvement in data sources on sales,
   prices, grants etc.

   6 *Identify*, from the relationships and behaviours, patterns of deprivation and opportunity for further selective analysis.

Such an examination of process and allocation is politically sensitive and may have no precise output. Furthermore, the model building is difficult. But concentration on process does allow for an understanding of the redistributive effects of policies and can set priorities. As a first step more rigorous models of local demand are needed based on improved data sources. A 'numbers game' based on demographic forecasting techniques is, however, more likely to continue to appeal to politicians and planners alike because of its simple, though inaccurate, application and output.

3.7 Conclusions

Housing forecasting is an inexact procedure. Whereas both housing need and demand are linked to population and employment change, the widely used procedure is, nevertheless, demographically based, with population estimates used as inputs in housing forecasts. Clearly, it is essential to integrate population, housing, employment and other forecasts to ensure feedback and consistency. Integration is also necessary from area to area, but no standard approach exists to ensure such integration in practice (see Chapter 8).

   Housing forecasts are almost always based on an accounting identity between forecast households and forecast housing stock. Demand forecasts are inadequately developed. Population figures, while a necessary input, should not form the basis for these analyses. Forecasts should be seen rather in the context of policy alternatives and consider a wide range of other factors including, for example:

1 Trends in employment, population, household formation, labour demand and migration.
2 Land availability.
3 Changes in existing stock, including improvement, rehabilitation, modernization and conversions.
4 Prices, rent levels, rent rebates and other aspects of housing finance.
5 The role of the local authority — passive or active.
6 Trends in the private rented sector.
7 Housing associations and voluntary groups.
8 The special needs of particularly vulnerable groups such as the old, disabled and otherwise disadvantaged.
9 Political pressures.

Factors such as these will influence both the aggregate level of housing provision and the tenure split, and it is not possible to separate out an objective method of housing forecasting from them.

4 Employment

4.1 Introduction

An appreciation and understanding of local economic activity provides an essential underpinning to much of the planner's work. For example, employment opportunities influence population levels, mainly through migration and, therefore, impact on housing requirements. Planners have an interest in three broad aspects of economic activity: local economic prospects and the consequent scale and character of development; personal incomes and the effect on demand for services such as shopping; and requirements for land, buildings and infrastructure. There are various approaches to the analysis of such activity, most having been developed within the field of regional science. Although the techniques to be considered in this chapter are applicable to most spatial scales, their use for local areas is not unproblematic. Because local economic activity is so strongly dependent on spatial interaction, meaningful analysis requires a definition of study boundaries which specifically acknowledges the interconnectedness of various activities/communities and, for this reason, the level of analysis is usually the region or urban area.

Not surprisingly, the regional scientist's tool kit relies very heavily on the theories and analytical techniques first developed for the study of national economies. But it is important to draw a distinction and highlight the divergence between the study of national economics and similar studies at a subnational scale. For example, trade between nations is constrained by a number of factors including exchange-rate fluctuations, tariffs, quota arrangements, and other physical and geographical barriers. With few such impediments at the subnational scale, regions and cities tend to specialize and trade to a much greater degree than do nations. Furthermore, when determining national economic strategy, policy makers can take advantage of a whole range of institutional tools to promote economic activity, for example, a currency devaluation, which are not available to planners at the subnational level. It is important to draw these distinctions because, while it is true to suggest that the national environment should always be considered in the formulation and implementation of local employment policy, it does not always follow that what is good for the nation is also good for all of its cities and regions.

A number of methods have been developed to improve understanding, and to facilitate the analysis, of local economic activity. On the one hand, planners have focussed on factors affecting urban location by developing theories of spatial structure to explain the nature and role of large urban agglomerations. More recently, however, the concern has switched from the location of activities, which is taken as given, to the relationship between activities and their impact on urban areas and their hinterlands. It is this second approach which provides the basis for the techniques developed to explain economic activity and helps us to take a forward view in the labour market. Within this latter category, this chapter will illustrate three of the more common techniques in current use and will also consider briefly a fourth, simpler approach based on what might be termed crude manpower forecasting. The first to be considered is economic base analysis which establishes a suitable framework for the subsequent consideration of other methods.

4.2 Economic base analysis

Underlying economic base analysis is economic base theory, which postulates that export industries provide the reason for the existence and growth of regions, cities or local areas.

4.2.1 Economic base theory

The more a given area specializes, the more it limits its self-sufficiency. Growth therefore depends on its ability to export goods and services to pay for its imported needs. Such exports are known as *basic* activity and are produced in the *exogenous* sector of the area's economy. However, numerous supporting activities are necessary to service the workers in these basic industries, their families and, of course, the industries themselves. These supporting activities, involving the output of goods and services for distribution and consumption solely within the given local area, are known as *non-basic* activities, the latter forming the *endogenous* sector of the area's economy. Basic activities, then, provide a means of payment for the goods and services a local area cannot produce for itself — for example, imported foodstuffs — and also support the services provided by non-basic industries, since the demand for the products of the latter comes from employees in the basic sector. The level of economic activity within a given area, however it is measured, is thus the sum of the levels in these two sectors. For example, if the unit of measurement is employment,

$$E = B + N \tag{4.1}$$

where E = total employment
B = basic employment
N = non-basic employment.

It can be seen, therefore, that although the exogenous and endogenous sectors are distinguished spatially in terms of the location of their demand areas, both sectors are, nevertheless, related to exogenous demand — the basic sector directly, and the non-basic sector indirectly by supporting the basic sector. Thus, given an increase in exogenous demand for the exports of a local area, the basic sector expands. This in turn generates an expansion in the supporting activities of the non-basic sector and, indeed, ultimately in population. The extent of the overall change is a multiple of the initial injection of basic employment. Growth, therefore, depends upon the response of the basic industries in the area to increased demand for their products from outside.

4.2.2 Empirical regularities

The ratio of non-basic employment to basic employment is called the *economic base ratio*. For example, if, in a particular study area, for every basic worker there are two non-basic workers, then the base ratio is 1:2; that is, the injection of an additional basic job leads to the creation of two further jobs in the non-basic sector (in the same way the loss of a basic job leads to a contraction). Given that the base ratio is 1:2, the *economic base multiplier* is 3, i.e., when basic employment increases by 1 a total of 3 new jobs are, in fact, created. Sometimes planners are interested in the relationship between levels of economic activity and population. At any one time a given number of jobs will support a certain number of people and the number of people supported by each job is called the *population multiplier*. Once the underlying social, economic and technological conditions/structure of a country have stabilized, the theory assumes certain empirical regularities, namely that the basic:non-basic ratios of local areas (cities, regions, etc.) and the ratios of these activities to total population remain constant. Arithmetically these relationships can be expressed as follows:

$$m = \frac{E}{B} \tag{4.2}$$

$$\alpha = \frac{P}{E} \tag{4.3}$$

where m = base multiplier B = basic employment
α = population multiplier P = total population
E = total employment

As this chapter is more concerned with employment change than population change, the focus of attention is on equation (4.2) above. For example, in a predictive study, given a change in the basic sector, the total employment impact on the local economy can be estimated by multiplying the change in basic activity by the base multiplier, i.e.

$$E = mB \tag{4.4}$$

4.2.3 The economic base study

When making forecasts of employment change in a local area, the actual economic base study is, therefore, relatively straightforward. First we need to identify the industries or categories in the basic and non-basic sectors. Existing employment in each sector can then be calculated and the base ratio and base multiplier computed using equations (4.1) and (4.2) above. Now given:

$$E_{(t)} = mB_{(t)} \tag{4.5}$$

then

$$E_{(t + n)} = mB_{(t + n)} \tag{4.6}$$

where E = total employment
         B = basic employment
         m = base multiplier
         t = base year
         $t + n$ = forecast year.

But total basic employment in the forecast year is the sum of basic employment in each of the categories (industries) which make up the basic sector. If we assume that there are k industries in the basic sector, then we can write

$$B_{(t + n)} = \sum_{i = 1}^{k} B_{i(t + n)} \tag{4.7}$$

where B_i = basic employment in category i.

Forecast year basic employment in each category can now be calculated by applying an appropriate growth factor to base year employment in that category. Growth factors are simply derived by compounding for the number of years in the forecast period — n years in this case — the percentage annual employment growth, where the latter is calculated for each category from historic data (as in trend lines in Chapter 2). Hence

$$B_{i(t + n)} = g_i B_{i(t)} \tag{4.8}$$

where g_i = a constant growth factor for category i.

Using equation (4.8) and substituting for $B_{i(t + n)}$ in equation (4.7):

$$B_{(t + n)} = \sum_{i = 1}^{k} g_i B_{i(t)} \tag{4.9}$$

The actual multiplier is derived, using equation (4.5), from base year totals for basic and total employment, that is:

$$m = \frac{E_{(t)}}{B_{(t)}} \tag{4.10}$$

Therefore, given equations (4.9) and (4.10), we can substitute values for m and $B_{(t + n)}$ into equation (4.6) as follows:

$$\begin{aligned} E_{(t + n)} &= m B_{(t + n)} \\ &= \frac{E_{(t)}}{B_{(t)}} \sum_{i=1}^{k} g_i B_{i(t)} \end{aligned} \tag{4.11}$$

Non-basic employment in the forecast year is simply obtained by subtracting basic employment from total employment, that is:

$$N_{(t + n)} = E_{(t + n)} - B_{(t + n)} \tag{4.12}$$

The predicted figure for non-basic employment can then be disaggregated, albeit crudely, by applying the base year proportions of the non-basic categories to the total, that is:

$$N_{i(t + n)} = N_{(t + n)} \frac{N_{i(t)}}{N_{(t)}} \tag{4.13}$$

where N = non-basic employment
N_i = non-basic employment in category i.

To illustrate the forecasting procedures in practice, assume a simple economy in which the basic sector comprises three industries (categories a, b and c). Base year employment totals have been obtained for each industry, as have historic growth rates. Given a 10-year forecasting period, Table 4.1 (overleaf) can be constructed. The non-basic sector of the economy also comprises three industries (categories d, e and f) and, once again, base year employment totals have been obtained for each—the data is in Table 4.2 (overleaf). As can be seen, the basic:non-basic ratio is 1:2 and the base multiplier is therefore 3, that is, given equation (4.5), then

cont.

$$m = \frac{E_{(t)}}{B_{(t)}}$$

$$= \frac{B_{(t)} + N_{(t)}}{B_{(t)}}$$

$$= \frac{450 + 900}{450}$$

$$= 3$$

Table 4.1 *Basic employment in the simple economy*

Industrial category	Existing basic employment in year t	Historic percentage annual employment growth	Growth factor for a 10-year forecast	Forecast basic employment in year t + 10
a	100	3	1.3	130
b	200	2	1.2	240
c	150	5	1.5	225
	$B_{(t)} = 450$			$B_{(t + 10)} = 595$

Using the computed value for the base multiplier and predicted changes in basic employment, the forecast for total employment is derived using equation (4.6) thus:

$$E_{(t + 10)} = mB_{(t + 10)}$$
$$= 3\,(595)$$
$$= 1785$$

Table 4.2 *Non-basic employment in the simple economy*

Industrial category	Existing non-basic employment in year t	Forecast non-basic employment in year t + 10
d	250	330.5
e	300	396.7
f	350	462.8
	$N_{(t)} = 900$	$N_{(t + 10)} = 1190$

The total for non-basic employment is therefore

$$N_{(t + 10)} = E_{(t + 10)} - B_{(t + 10)}$$
$$= 1785 - 595$$
$$= 1190$$

The forecast total for non-basic employment can now be disaggregated to complete the final column of Table 4.2 using equation (4.13) as follows:

$$N_{d(t + 10)} = N_{(t + 10)} \frac{N_d(t)}{N(t)}$$
$$= 1190 \times \frac{250}{900}$$
$$= 330.5$$
$$N_{e(t + 10)} = 1190 \times \frac{300}{900}$$
$$= 396.7$$
$$N_{f(t + 10)} = 1190 \times \frac{350}{900}$$
$$= 462.8$$

4.2.4 The limitations of the economic base approach

Although the use of the economic base technique appears relatively straight-forward, many criticisms can be expressed regarding the validity of the approach. First there is the unit of measurement: employment data is used because it is easier to obtain and to calculate the multipliers, but this ignores wage levels. If real wages in the basic sector rise, then demand and (prob-ably) employment will rise in the non-basic sector. Secondly, there are considerable difficulties in actually distinguishing between basic and non-basic employment (see Massey, 1973). Indeed as the area of study decreases in size, so the proportion of basic industry increases as proportionately more product is 'exported'. However, assuming that it is possible to differentiate between the two sectors and that appropriate units of measurement can be found, another severe problem arises: namely, there is still unlikely to be a constant basic:non-basic ratio for the study area even if the underlying economic conditions remain fairly stable. A number of factors can alter this ratio, not least the differing economies of scale likely to be experienced in the two sectors as the economy grows. This naturally leads to a fourth area of concern, namely the extent to which it can reasonably be assumed that there is stability over the forecasting period in the values of the growth indices and the multiplier. A fifth criticism highlighted by a number of commentators relates to the technique's tendency to underestimate the importance of the non-basic sector in promoting economic activity — a well developed non-basic sector can act as a powerful stimulant to growth because of its ability to attract basic industry. But perhaps the most important and fundamental flaw in economic base analysis is its almost total neglect of what economists refer to as supply-side limitations.

The economic base model is a demand-orientated model which omits the supply side of the labour market as one of its simplifying assumptions. This can lead to considerable ambiguity which arises because of the failure of the analysis explicitly to recognize cause-and-effect modes of explanation. This is an issue raised by Sayer (1976) who emphasizes the need to 'distinguish between *ex ante* quantities (referring to market expectations and intentions) and *ex post* quantities (referring to actual, realized quantities).' When this distinction is made, important changes that might occur in the underlying conditions of supply and demand are highlighted despite the apparent stability of the important relationships so fundamental to the model. The point is best illustrated with an example. Assume that there is a proposal to provide 100 additional basic jobs in a particular area where the base multiplier is 3. The economic base model assumes a one-way causation from basic jobs to total employment and predicts an increase in total employment, subject of course to an appropriate time lag, of 300. But these vacancies may or may not be taken up depending on supply conditions. There may only be 50 workers with the appropriate skills in (or able to migrate to) our local area able to fill the vacancies. If only half the jobs are taken up therefore, and the final result is an increase in total employment of 150, the validity of the model is apparently maintained as an *ex post* relationship cleverly concealing the unsatisfied demand. Analysts obviously need to be conscious of this 'illusion' when using the technique.

Notwithstanding the extent to which the above criticisms limit the effectiveness of the approach, if used constructively economic base analysis has, nevertheless, proved to be a highly utilitarian tool for exploring changes in economic activity. It is relatively simple both conceptually and operationally and, assuming that economic base theory provides a reasonable explanation for the growth of the local area under study, lends itself to the kind of rough impact/trend assessments that are so useful to planners in preliminary analytical studies. It can and has, of course, been refined in a number of studies to provide more detailed guidance for development plans, but it is an aggregate technique and this neglect of inter-industry relationships remains a major criticism. A better perspective would be provided by a disaggregate approach which not only looked at the performance of individual industries, but also the interrelationships between these industries. This additional dimension is provided by input–output analysis, which demonstrates and quantifies the relationships that exist between the various sectors of the economic system but, like the economic base method, can also be used to measure the relationships between a given study area and its hinterland. Indeed, it has been suggested that input–output analysis is simply a form of economic base analysis taken to the limits of its possibilities (Bendavid-Val, 1983).

4.3 Input–output analysis

As the name implies, input–output analysis is concerned with the inputs and outputs of each industry within a particular study area and their inter-relationships. The outputs of any industry can either be consumed, because they are what economists refer to as final goods, or they are intermediate goods to be used as industrial inputs in further productive activity. The input–output method traces the flow of these outputs of goods and services (measured in either physical or monetary terms) from one sector of the economy to another.

4.3.1 Inter-industry accounting

Input–output analysis is often referred to as inter-industry accounting. Because it deals explicitly with the inter-industry transactions that arise from the demand for final products, it differs from other forms of accounting which exclude these transactions for intermediate goods to avoid double counting.

The analysis begins by separating the economy into producers and consumers and then focusses on the extent to which these are engaged in final or intermediate transactions. Producers, for example, are either *intermediate producers/suppliers* purchasing inputs which are then sold to other intermediate producers or for final consumption, or *primary producers/suppliers* who do not purchase inputs to make what they sell. Payments to the latter do not generate further inter-industry sales and the earnings of primary producers, therefore, represent value added. Consumers can also be either *intermediate purchasers*, buying the output of products for further processing, or *final purchasers*, buying finished products for final consumption. Clearly intermediate producers and consumers are the same, whereas primary producers and final consumers may not be.

In summary, input–output analysis considers an economic system as an aggregate of mutually related industries. Each industry takes inputs, such as raw materials, from other industries which it transforms into outputs and subsequently supplies as input for other industries in the economy. Since production to satisfy final demand generates a whole series of such intermediate transactions, the final output of an input–output study usually takes the form of a series of matrices in which the output of each industry is recorded and represents input for other industries in the economy. These matrices identify the nature and quantify the degree of structural inter-dependence between the various components in the system. If final demand changes, the input–output model can then be used to highlight the ramifications by identifying the necessary production changes of the industries back down the line. These concepts will become more clear in a discussion of the individual components that make up the input–output approach.

4.3.2 The three input–output matrices

The input–output method employs three matrices. The *transactions matrix* describes the flow of goods and services between buyers and sellers. These flows, which can be measured in either quantitative or value terms, are usually measured in terms of money and viewed as sales transactions. There are then two technology matrices which summarize the economy's technological conditions. The first, an *input-coefficient matrix*, is derived from the transactions matrix and shows the inputs required from suppliers by each immediate purchaser per unit of output that it produces. This is sometimes known as the *direct-requirements table* and can be used to derive the second of our technology matrices, i.e. the *total-requirements table*, which shows the total purchases of direct and indirect inputs that are required by intermediate producers per unit of output sold to final purchasers.

(a)   The transactions matrix

Assume an economy is composed of *n* industries each producing only one type of product. The output of each industry is then distributed to other industries as input or to households for final consumption. The productive activity of an industry employing primary inputs represents the value added as part of the industry's output. Now using general notation, if x_{ij} represents the output of industry i used by industry j, u_i the final demand from industry

Table 4.3   *Generalized transactions matrix*

Output \\ Input		Intermediate purchasers 1 2 $\cdots$ n	Final purchasers	Total sales (outputs)
Intermediate suppliers	1 2 . . . n	x_{11} x_{12} $\cdots$ x_{1n} x_{21} x_{22} $\cdots$ x_{2n} . . . . x_{n1} x_{n2} $\cdots$ x_{nn}	u_1 u_2 . . . u_n	X_1 X_2 . . . X_n
Primary suppliers		p_1 p_2 $\cdots$ p_n	u_{n+1}	
Total purchases (inputs)		X_1 X_2 $\cdots$ X_n		

i, X_i the total sales of industry i, p_j the primary input of industry j, and X_j the total purchases of industry j, then the transactions for the whole economy can be tabulated as in Table 4.3. This is the transactions matrix in which the rows show the distribution of each supplier's sales to intermediate and final purchasers, and the columns show the distribution of each purchaser's purchases from intermediate and primary suppliers. The total sales X_i of industry i is therefore

$$X_i = \sum_{j=1}^{n} x_{ij} + u_i \tag{4.14}$$

while the total purchase X_j of industry j is

$$X_j = \sum_{i=1}^{n} x_{ij} + p_j \tag{4.15}$$

and in general

$$X_i = X_j$$

The overall process is best considered with a numerical example. Consider a simple economy in which all industries have been reduced to just two sectors, commercial and manufacturing. Furthermore, there is only one final purchaser which will be called households and also one primary supplier, once again households, supplying primary inputs such as labour and entrepreneurship. Finally, this hypothetical economy does not trade with the outside world so there are no imports and exports, there are no savings or investment, and there is no government intervention. A typical transactions matrix for such an economy is illustrated in Table 4.4. Assuming that all transactions are measured in monetary terms, the first row of data shows that manufacturing's total sales was £300. Of this £300, manufacturing sold £150 of output to itself for further processing, £50 to the commercial sector for further processing and £100 to households for final consumption. Similarly, the first column shows that in order to produce £300 of total output, manufacturing had to purchase £150 of products from itself, £100 from the commercial sector, and £50 in primary inputs from households. In the same way the commercial rows and columns can be analysed. Readers should note that for any intermediate industry, total inputs equal total outputs and, for the economy as a whole, total final purchases equal total primary inputs. Hence the south-east corner of the matrix shows that total inputs into the system will equal total outputs of the system.

Although the transactions matrix provides a very comprehensive picture of the inter-industry flows of goods and services in the economy during any

Table 4.4   *Transactions matrix for the two-sector economy*

Output / Input	Intermediate purchasers		Final purchasers	Total sales
	Manufacturing	Commercial	Households	
Intermediate suppliers:				
Manufacturing	150	50	100	300
Commercial	100	75	75	250
Primary suppliers:				
Households	50	125	25	200
Total purchases	300	250	200	750

given study period, its role is essentially descriptive. The technology matrices are more useful analytical devices, the input-coefficient matrix in
particular contributing what Ayeni (1979) has referred to as the 'cooking
recipe' of each sector of the economy. It offers a technique for projecting
into the future the magnitude of important sectors and linkages in the
economy.

(b)   The input-coefficient matrix
The input-coefficient matrix describes the structure of the economy. Its
construction involves a transformation of the basic data in the transactions
matrix into a generalized statement of direct input requirements per unit of
output for each intermediate industry. These coefficients are derived by
dividing the input figures in each intermediate purchaser column by the
number at the bottom of the column, total inputs for that industry. Since
total inputs equals total outputs, dividing each intermediate purchaser's
column through by the total will provide a distribution of inputs per unit of
output for each intermediate supplier. In other words, using the general
notation, the coefficients are determined as follows:

$$a_{ij} = \frac{x_{ij}}{X_j} \quad \left(\text{or } \frac{x_{ij}}{X_i}\right) \quad \text{for } i, j = 1, 2, \ldots, n.$$

As before, the procedure is best illustrated with a numerical example, so
given Table 4.4, dividing each intermediate purchaser column through by

the total will provide the distribution of inputs per unit of output for each intermediate supplier. The results of this simple computation produces the input-coefficient matrix as illustrated in Table 4.5.

Table 4.5 *Input-coefficient matrix*

Output Input	Intermediate purchasers		Final purchasers
	Manufacturing	Commercial	Households
Intermediate suppliers:			
Manufacturing	0.5	0.20	0.50
Commercial	0.33	0.30	0.38
Primary suppliers:			
Households	0.17	0.50	0.13
Total purchases	1.00	1.00	1.00

The input-coefficient matrix provides the *direct requirements* necessary to calculate direct inputs for any level of demand for the output of any intermediate industry. For example, Table 4.5 shows that if the expected final demand for manufacturing is £1000, then satisfying this demand will require direct intermediate inputs to manufacturing of £500 of manufactured products (i.e. 1000×0.5), £330 of commercial products (i.e. 1000×0.33) and £170 of primary inputs from households (i.e. 1000×0.17). However, these are only first round inputs directly required to produce the final demand items that households want. The first round input requirements must be produced as well and they require a whole set of inputs too. The input requirements in this second round are called *indirect inputs* and they, in turn, will need to be produced, thus giving rise to a third round of indirect inputs. Theoretically, this iterative round-by-round computation is continued *ad infinitum*. By summing the input requirements of all rounds by type of input, we obtain the output of each sector required both directly and indirectly to yield the set of sales to final purchasers. In practice, an infinite number of rounds need not be computed, because the rounds of input requirements become smaller and smaller, so that after the fifth or sixth round the sum of successive iterations can be approximated with a sufficient degree of accuracy. This is where the input–output approach can be used for projection rather than description.

Table 4.6 Total requirements computation for the simple economy derived from the input-coefficient matrix

| | | Sales as direct inputs | | | Sales as indirect inputs | | | | | | | | | | | | | |
	Sales to final purchasers	First round			Second round			Third round			Fourth round			Fifth round			Total	Total Sales
		To man.	To comm.	Total	To man.	To comm.	Total	To man.	To comm.	Total	To man.	To comm.	Total	To man.	To comm.	Total	Total	Total Sales
By man.	200	(0.5)(200) =100	(0.2)(100) =20	120	(0.5)(120) =60	(0.2)(96) =19.2	79.2	(0.5)(792) =39.6	(0.2)(68.4) =13.68	53.28	(0.5)(53.22) =26.64	(0.2)(46.66) =9.33	35.96	(0.5)(35.97) =17.99	(0.2)(31.58) =6.32	24.31	192.8	512.8 +
By comm.	100	(0.33)(200) =66	(0.3)(100) =30	96	(0.33)(120) =39.6	(0.3)(96) =28.8	68.4	(0.33)(79.2) =26.14	(0.3)(68.4) =20.52	46.66	(0.33)(53.28)(0.3)(46.66) =17.58	=14	31.58	(0.73)(35.97)(0.3)(31.58) =11.27	=9.47	21.34	162	364
By households	—	(0.17)(200) =34	(0.5)(100) =50	84	(0.17)(120) =20.4	(0.5)(96) =48	68.4	(0.17)(79.2) =13.46	(0.5)(68.7) =34.2	47.66	(0.17)(53.22)(0.5)(46.66) =9.06	=23.33	32.39	(0.17)(35.97)(0.5)(31.52) =6.11	=15.79	21.9	170.4	255.4 +
Subtotal				300			216			167.6			99.94			67.55		
By all suppliers	300																531.1	113.1 +

Table 4.7 Computation of total-requirements coefficients for manufacturing sales of £1 to final purchase

| | | Sales to direct inputs | | | Sales to indirect inputs | | | | | | | | | | | | | |
	Sales to final purchasers	First round			Second round			Third round			Fourth round			Fifth round			Total	Total Sales
		To man.	To comm.	Total	To man.	To comm.	Total	To man.	To comm.	Total	To man.	To comm.	Total	To man.	To comm.	Total	Total	Total Sales
By man.	1.00 / 1.00	(0.5)(1) =0.5	—	0.5	(0.5)(0.5) =0.25	(0.2)(0.33) =0.066	0.316	(0.5)(0.316) =0.158	(0.2)(0.264) =0.053	0.211	(0.5)(0.211) =0.106	(0.2)(0.183) =0.037	0.143	(0.5)(0.143) =0.072	(0.2)(0.125) =0.025	0.077	0.767	2.267 +
By comm.	—	(0.33)(1) =0.33	—	0.33	(0.33)(0.5) =0.165	(0.3)(0.33) =0.099	0.264	(0.33)(0.316) =0.104	(0.3)(0.264) =0.079	0.183	(0.33)(0.211) =0.07	(0.3)(0.183) =0.055	0.125	(0.33)(0.143) =0.047	(0.3)(0.125) =0.038	0.085	0.657	0.987 +
By households	—	(0.17)(1) =0.17	—	0.17	(0.17)(0.5) =0.085	(0.5)(0.33) =0.165	0.25	(0.17)(0.316) =0.054	(0.5)(0.264) =0.132	0.186	(0.17)(0.211) =0.036	(0.5)(0.183) =0.092	0.128	(0.17)(0.143) =0.024	(0.5)(0.125) =0.063	0.087	(0.651)	(0.821) 0.83 1.00
By all suppliers	1.00			1.00													2.254	4.254 +

(c)   The total-requirements matrix
Assuming constant production coefficients (i.e. the cells in the input-coefficient matrix) and given some forecast or estimate of sales to final demand, the input-coefficient matrix can be used to derive a total-requirements matrix showing the total purchases of direct and indirect inputs that are required throughout the economy per unit of output sold to final purchasers by an intermediate supplier. For example, assume in our simple economy that sales to final purchasers at some future date are estimated to be £200 and £100 for manufacturing and commercial respectively. Table 4.6 illustrates the implications for the economy. If manufacturing sales to final purchasers are £200, multiplying this figure through the manufacturing column of the input-coefficient matrix shows that direct inputs of £100 are required from manufacturing, £66 from commerce and £34 from households as primary suppliers, that is column 2 in Table 4.6. The process is repeated to calculate the direct-input requirements to enable commerce to satisfy demand of £100 from final purchasers, that is, column 3 in Table 4.6. Total direct requirements to satisfy final demand is, therefore, as illustrated in column 4 in Table 4.6. Total direct-input sales can now be multiplied through by their respective columns in the input-coefficients matrix to arrive at second-round inputs (the first set of indirect requirements) and so on. As stated above, the procedure should continue for several iterations, the number depending on the particular problem and the degree of refinement required. However, in the illustration only five iterations have been completed — hence the plus signs added to the totals for indirect inputs to indicate that these are underestimates.

In the same way that a set of input coefficients was derived from the transactions matrix, it would clearly be useful to derive a set of *total-requirements coefficients* to provide a more generalized statement of the linkages within the economy. This can be done by simply applying the iterative method for £1 of sales to final purchasers for each intermediate industry in turn. Table 4.7 illustrates this process for manufacturing sales of £1 to final purchasers. Readers should note that two sets of numbers are listed for total sales by households. The bracketed figures are derived from the five iterations in the analysis. However, total sales by primary suppliers must, of necessity, equal final purchases. So setting them equal to 1.00, total indirect primary inputs can be easily calculated as the difference between final purchases and direct primary inputs (1.00 − 0.17 = 0.83). Hence the two numbers not in brackets, which are the ones used in calculating total requirements from all suppliers. In summary, the figures in the final column in Table 4.7 are our total-requirements coefficients and indicate the total requirements from all sectors of the economy to supply £1 of manufacturing's product to final purchasers. A similar set of coefficients could be derived for the commercial sector in our simple economy (and, indeed, for any other intermediate industries in more complex studies) to produce a total-requirements coefficient matrix. The total-requirements coefficients derived thus are similar to the economic base multipliers discussed earlier in the chapter. For example, given the total requirements coefficient for manufacturing of 4.254 as in Table 4.7, then for every £1 of exogenous demand that is satisfied by manufacturing, a total of £4.254 in sales will have been generated.

Given the complexities of real economies, the procedures as illustrated above would be quite cumbersome if they were to be performed manually. However, the round-by-round iterations can in effect be performed through the use of a matrix called the *Leontief inverse*, which lends itself to easy computation. The latter is derived from the input-coefficient matrix, and shows how much output must be produced in a given industrial sector in order that one unit of final demand sales can be made — the matrix yields both direct and indirect requirements. The actual use of matrices can be demonstrated with a simple example.

As discussed earlier, there are two types of demand — intermediate, inter-industry demand and the demand of final users. Taking the economy's transactions matrix and the input-coefficient matrix into consideration, and remembering how each cell in the latter is derived, that is, $a_{ij} = x_{ij}/X_j$ (or x_{ij}/X_i), the inter-industry demand can be written as AX, where

$$
A = \begin{bmatrix}
a_{11} & a_{12} & \cdots & a_{1n} \\
a_{21} & a_{22} & \cdots & a_{2n} \\
\cdot & \cdot & & \cdot \\
\cdot & \cdot & & \cdot \\
\cdot & \cdot & & \cdot \\
a_{n1} & a_{n2} & & a_{nn}
\end{bmatrix}
\qquad
X = \begin{bmatrix}
X_1 \\
X_2 \\
\cdot \\
\cdot \\
\cdot \\
X_n
\end{bmatrix}
$$

and the demand of final users U is given by

$$
U = \begin{bmatrix}
u_1 \\
u_2 \\
\cdot \\
\cdot \\
\cdot \\
u_n
\end{bmatrix}
$$

Within the economy, production must be adjusted to meet total demand, i.e.

$$X = AX + U$$
$$\text{or} \quad U = X - AX$$
$$= (I - A) X$$

This is a very basic description of the input–output model in which the quantities produced are assumed to be non-negative, i.e.

$$X_i, X_j, u_i \geqslant 0,$$

It is beyond the scope of an introductory text such as this to consider the conditions imposed on the input-coefficient matrix A in order to have a non-negative solution. The model as set out can now be used in predictive studies. More specifically, given the structure of the economy (matrix A) and the demand from final purchasers (vector U), it is necessary to know the outputs of all industries in the economy in order to fulfill all the demand requirements. If the matrix $(I - A)$ has an inverse (see Appendix A.12) then it is possible to solve for X thus:

$$X = (I - A)^{-1} U$$

To demonstrate numerically, consider the simple two-sector economy whose transactions matrix is shown in Table 4.4.

The problem is to determine the outputs of the two sectors if final use demand (final purchases) changes to:

$$U = \begin{bmatrix} 150 \\ 100 \end{bmatrix}$$

As explained earlier, from the transactions matrix the input matrix was found to be as shown in Table 4.5.

Thus
$$A = \begin{bmatrix} 0.50 & 0.20 \\ 0.33 & 0.30 \end{bmatrix}$$

Therefore
$$I - A = \begin{bmatrix} 1-0.50 & -0.20 \\ -0.33 & 1-0.30 \end{bmatrix} = \begin{bmatrix} 0.50 & -0.20 \\ -0.33 & 0.70 \end{bmatrix}$$

And:
$$\begin{aligned} |I - A| &= (0.50)(0.70) - (-0.33)(-0.20) \\ &= 0.284 \end{aligned}$$

Therefore:
$$(I - A)^{-1} = \frac{1}{0.284} \begin{bmatrix} 0.70 & 0.20 \\ 0.33 & 0.50 \end{bmatrix}$$

Now if
$$U = \begin{bmatrix} 150 \\ 100 \end{bmatrix}$$

and
$$X = (I - A)^{-1} U$$

then
$$\begin{aligned} X &= \frac{1}{0.284} \begin{bmatrix} 0.70 & 0.20 \\ 0.33 & 0.50 \end{bmatrix} \begin{bmatrix} 150 \\ 100 \end{bmatrix} \\ &= \begin{bmatrix} 440.14 \\ 350.35 \end{bmatrix} \end{aligned}$$

Therefore, in order to satisfy final user demand of $\begin{bmatrix} 150 \\ 100 \end{bmatrix}$, total outputs of £440.14 must be produced by manufacturing and £350.35 by commerce. Obviously there are far more complicated economic and mathematical problems involved in input–output analysis than those suggested by the above example, and readers are referred to Leontief (1966).

The input–output model is clearly a concise way of studying a local economy. However, if the simple analysis is extended and the possibility of trade between regions is introduced, the approach is made considerably more complex because interest is now in knowing the sources of final purchases and primary supplies, and in distinguishing between the regional/ local economy and the rest of the world. The general methodology is, nevertheless, the same, although readers should note that when trade is introduced, intermediate suppliers and purchasers are those located within the study area — imported supplies (whether processed or not) are not intermediate inputs as far as the local economy is concerned because their production does not require locally produced inputs, and export sales (whether for intermediate use or final consumption) are not intermediate sales for the local economy because they are not processed further within the study area.

4.3.3 The limitations of input–output analysis
Apart from the obvious operational problems that are likely to be encountered when undertaking an input–output study, not least the formidable data assemblage for the construction of the transactions matrix, a number of other limitations of the approach need to be addressed.

One of the most important criticisms of the input–output model when used in predictive studies is the assumption of the stability of the technological coefficients. In reality, these are known to vary with time and, as a generalization, it can be said that the greater the industry detail in the input–output tables and the rate of innovation in the study area, the less reliable will be the input coefficients used for long-term analysis. Leontief (1966) and others have attempted to overcome this obvious deficiency by developing a dynamic theory of input–output systems. However, the techniques used when introducing dynamic considerations require an advanced knowledge of calculus and difference equations beyond the scope of this text.

A second important limitation with the basic input–output technique and, indeed, with its dynamic formulation, is that both are essentially linear expressions which assume that the effect of undertaking several types of production is the sum of the separate parts. Unfortunately, this additivity assumption ignores the possibility of both external economies and diseconomies in the production process.

A third problem arises with prices since actual sales data is often either incomplete or, sometimes, not available. In such cases, estimates need to be made, but this is not a trivial exercise because most goods sell at one price to intermediate purchasers and at another for final consumption. The pricing issue is particularly important, not only because transactions data provides the basis for calculating the technical coefficients but, in the context of this chapter, because accurate prices often provide the basis for converting the results of the input–output study into labour and hence employment terms.

Finally, in predictive studies, the use of input–output analysis requires independent final sales estimates and the reliability of projections will obviously depend on how speculative these estimates are.

Despite its obvious shortcomings in predictive studies, there is considerable scope for exploiting the use of input–output analysis as a descriptive tool. By providing a comprehensive picture of the linkages within the local economy it highlights the strategic importance of various industries and sectors. It is not, however, a technique that can be easily used to analyse the nature of such change and, to this end, attention is now turned to shift-and-share analysis.

4.4 Shift-and-share analysis

Shift-and-share analysis is a relatively uncomplicated technique that attempts to identify some of the factors that underlie the differences in the growth and hence employment performance of different regions or local areas. In recognizing that different areas will grow at different rates, the approach simply asks why this should be the case.

4.4.1 The competitive and industry-mix effects

There are two main reasons why a region or local area may grow at a slower or faster rate than the national average. First, different areas possess different attractions to and conditions for industrial development and hence it would be expected that there would be different growth rates in similar industries. The second component recognizes that, within particular areas, the mix of industries might be strongly weighted towards the slow or fast growth type — for example, areas with high concentrations of industries with a strong propensity to grow would be expected to grow faster than other areas within the national or macro-regional space economy. Clearly, it would be very useful if planners could isolate these two effects, known as the competitive effect and the industry-mix effect, from the national growth trend, and to identify and quantify the extent to which each is responsible for a given study area's growth performance. This procedure of separating the total shift into its two major components is the shift-and-share analysis.

4.4.2 The shift-and-share method

The actual operationalization of the technique is based, as suggested above, on a comparison of the growth performance of a given local area or region with the expected growth in a larger area (macro-region or nation) of which the study area is just one of many. National or macro-regional and study area data — measured either in terms of output, value-added, or employment — are thus required for two periods in order to proceed with estimation as follows.

Assume that employment data is being used and that $E_{ij(t)}$ is the level of employment in sector i and in region or local area j at time t. If there are m industries and l regions or local areas, the regional or local area employment change in each industry is then

$$\Delta E_{ij} = E_{ij(t + n)} - E_{ij(t)} \qquad (4.16)$$

where ΔE_{ij} = change in employment between time periods t and $t + n$.
    n = the number of years between the two periods.

At the national or macro-regional scale, employment change in each industry, ΔE_{iN}, is

$$\Delta E_{iN} = \sum_{j=1}^{l} \Delta E_{ij} \qquad (4.17)$$

While national or macro-regional employment change in all activities, ΔE_N, is

$$\Delta E_N = \sum_{i=1}^{m} \sum_{j=1}^{l} \Delta E_{ij} \qquad (4.18)$$

Now if we assume that all regions or local areas grow at the national or macro-regional rate, then the industry's hypothetical growth is $(\Delta E_N / E_{N(t)}) E_{ij(t)}$. Subtracting this from the growth that actually occurred, ΔE_{ij}, a measure is obtained than can be called the industry's total shift, S_{ij}, that is

$$S_{ij} = \Delta E_{ij} - \left[\frac{\Delta E_N}{E_{N(t)}} \right] E_{ij(t)} \qquad (4.19)$$

where S_{ij} = the net shift in industry i and in region j of the l region space economy.

There is now a need to determine the magnitude of the major forces contributing to this net shift. As stated earlier, the first component is a competitive effect commonly known as the *differential shift*, and measures the study area's locational advantages. If this is defined as D_{ij}, then

$$D_{ij} = \left[\frac{\Delta E_{ij}}{E_{ij(t)}} - \frac{\Delta E_{iN}}{E_{iN(t)}} \right] E_{ij(t)} \tag{4.20}$$

As can be seen, the size of the differential shift depends on the difference between the industry's regional and local area growth rate and its national or macro-regional growth rate. It measures the extent to which forces operating on industry at the regional level (e.g. labour supply advantages) contributed to the magnitude of the net total shift. It is possible, therefore, for equation (4.20) to be either positive or negative, the former indicating that the study area has a competitive advantage, the latter otherwise.

The second component of the total net shift focusses on industrial composition and measures the effect of the study area's industry mix. This is commonly known as the *proportionality shift* and, if this is defined as P_{ij}, then

$$P_{ij} = \left[\frac{\Delta E_{iN}}{E_{iN(t)}} - \frac{\Delta E_{N}}{E_{N(t)}} \right] E_{ij(t)} \tag{4.21}$$

The size of the proportionality shift depends on the difference between the total national growth rate and the industry's national growth rate. It therefore measures the extent to which forces operating on the industry at the national level (e.g. changes in national demand patterns) contributed to the magnitude of the total net shift. If the proportionality shift is positive, it means that the study area has a favourable industry-mix biassed towards rapid growth industries.

That the differential shift and the proportionality shift together equal the total shift is easily verified by simply adding them, that is:

$$D_{ij} + P_{ij} = \left[\frac{\Delta E_{ij}}{E_{ij(t)}} - \frac{\Delta E_{iN}}{E_{iN(t)}} \right] E_{ij(t)} + \left[\frac{\Delta E_{iN}}{E_{iN(t)}} - \frac{\Delta E_{N}}{E_{N(t)}} \right] E_{ij(t)}$$

$$= \left[\frac{\Delta E_{ij}}{E_{ij(t)}} - \frac{\Delta E_{N}}{E_{N(t)}} \right] E_{ij(t)}$$

$$= \Delta E_{ij} - \left[\frac{\Delta E_{N}}{E_{N(t)}} \right] E_{ij(t)}$$

$$= S_{ij}$$

The approach as described above, which is illustrated schematically in Figure 4.1, can be adapted for predictive purposes by assuming that regional shifts remain constant and incorporating a differential growth rate — this is

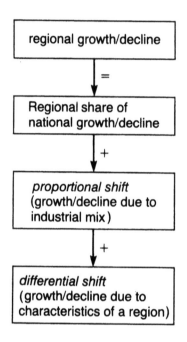

Figure 4.1 *Shift-share analysis.*

the so-called *constant shift method.* Beginning with equation (4.20) which expresses D_{ij} in terms of differential rates of growth it is possible to proceed as follows:

Given

$$D_{ij} = \left[\frac{\Delta E_{ij}}{E_{ij(t)}} - \frac{\Delta E_{iN}}{E_{iN(t)}} \right] E_{ij(t)}$$

then

$$\frac{D_{ij}}{E_{ij(t)}} = \frac{\Delta E_{ij}}{E_{ij(t)}} - \frac{\Delta E_{iN}}{E_{iN(t)}} \tag{4.22}$$

This gives the differential growth rate that industry *i* experienced in study area *j*, i.e. it is the difference between industry *i*'s recent study area and national or macro-regional growth rates. Equation (4.22) may be written for ΔE_{ij} as

$$\Delta E_{ij} = \left[\frac{\Delta E_{iN}}{E_{iN(t)}} + \frac{\Delta D_{ij}}{E_{ij(t)}} \right] E_{ij(t)} \tag{4.23}$$

Equation (4.23) is the projection equation under the constant shift assumption. As can be seen, the national growth-rate term has been relocated to the same side as the differential growth-rate term, and the technique therefore takes account of the possibility that regional shares may shift and that such changes may be important. For example, if it is found from historic data, say the last five years, that the study area's manufacturing sector had a growth rate that exceeded the national (or macro-regional) growth rate for manufacturing by 2.3% and, assuming that there is a five-year national (or macro-regional) growth rate projection for manufacturing of 2.7%, then the study area's recent differential growth rate is added to the national/macro-regional growth rate projection to obtain the projected study area growth rate, i.e. 2.3% + 2.7% = 5%. This result can now be applied to the study area's current manufacturing output, value added or employment levels (depending on which units we are working in) to obtain our five-year growth projection.

4.4.3 The limitations of shift-and-share analysis
A first and obvious criticism of the technique is that it is essentially descriptive. It shows what differences exist between areas and how these might change, but it does not explain why. It does not explain why a particular industry mix exists, why different industries experience different growth rates and so on. Secondly, the results are highly sensitive to the degree of disaggregation. The finer the level of classification of industrial structure, the larger is the proportional shift and the smaller the differential shift. Yet there is no theoretical justification for any particular level of disaggregation. In common with the other techniques considered in this chapter, shift-and-share analysis also suffers because of its failure adequately to embrace the consequences of technical progress. This is a particularly acute problem when using employment as the unit of measurement since, as Bendavid-Val (1983) points out, it 'results in a systematic underestimate of the overall growth impact of industries undergoing the most rapid labour productivity gains.' A further criticism is Brown's (1969) finding that the industry-specific differential growth rates as measured by equation (4.22) are unstable and, as a result, the differential growth rates of the recent past are an unreliable proxy for the projected differential growth rates used in forecasting.

Another problem arises in that industry interdependence may underestimate the proportionality shift. A below average representation of growth industries means less auxiliary services, and so part of a negative differential shift may be traced back to industrial structure. Put simply, shift-and-share analysis may underestimate the influence of industrial structure.

The problems of the technique have prompted Richardson to dismiss it as '... perhaps the most common and most overvalued tool of analysis in regional economics ... It is ... a harmless pastime for small boys with pocket calculators.' (Richardson, 1978, p. 202). Such damning criticism suggests caution. Shift-and-share analysis has, nevertheless, something to commend it, at least because of the clarity with which it highlights the relationships and differences between the study area's and national economic performance. It provides an overview of the role of the study area in the broader national or macro-regional industrial complex and, when used in conjunction with other techniques, it may be a useful aid to decision makers. Readers should also note that the description of the technique in this chapter is capable of refinement and further development (see, for example, Krueckeberg and Silvers, 1974).

4.5 Crude forecasts

Not only do the techniques outlined above involve substantial work and extensive data requirements but, as has been indicated, they are subject to important theoretical and technical limitations which, at the local level, are even more pronounced. It is little wonder that local authority planners have opted for much simpler techniques, with approaches based on the kind of crude forecasting methods employed by manpower planners at the national level finding much favour of late.

Manpower forecasting involves estimating for some future time period the number of people who will be available and looking for work, and the number of jobs employers will want to fill. These estimates of labour supply and demand are normally made independently and set against one another, though in reality availability of employees and the nature of employer's requirements are interdependent. SDD Advice Note No. 4 (Scottish Development Department, 1975), suggests the following method for structure plans:

Calculate labour supply
  (a) Calculate male and female activity rates in the study area and nationally for population aged 16+.
  (b) Calculate the ratio of the study area and national rates.
  (c) Subjectively modify for future.
  (d) Apply future ratio to the national forecast to give a study area forecast.
  (e) Forecast population and apply activity rates.
  (f) Subtract forecast unemployment.
  (g) Adjust for travel-to-work areas where these differ from the local authority study area.

Calculate labour demand

(a) Collect disaggregated employment data.
(b) Collect employment data for major employers.
(c) Collect (a) and (b) above by journey-to-work areas within the study area.
(d) Collect (a) and (b) for males and females separately.
(e) Collect the above data for a past period as long as the forecast period.
(f) Extrapolate employment figures taking into account employers' opinions.
(g) Consider major employers separately.
(h) Adjust 'to ensure sense'.
(i) Adjust for cyclical trends.

It then suggests careful monitoring with annual reviews, and also emphasizes the need to link the forecasts to national forecasts.

The substantial problems of the approach are identified, for example, as already stated, the supply of and demand for labour are interrelated and not independent. Furthermore, whereas some analysts have suggested that forecasts of labour supply are not too difficult to make and that demand forecasts are much more problematic, labour demand assumptions may be input into population forecasts from which labour supply is calculated. Nonetheless, this is the recommended method. At first it seems excessively simplistic, if not naïve. But perhaps it is merely a recognition of the difficulty in applying more elaborate techniques at the local level.

4.6 Conclusions

The need for a forward view in local labour markets is not in doubt, but an important question arises as to how accurate and precise these forecasts can be. Even at the national and macro-regional scale, when predicting aggregate changes in the general level of economic activity, recent evidence suggests that the forecasting process is open to wide margins of error. And yet analysts have proceeded optimistically with apparently more detailed forecasts disaggregated both sectorally and spatially. Does not the variety and complexity of the individual market decisions, made by employers and employees alike, make forecasting in such detail a vain panacea? Not necessarily. The geographical concentration of unemployment highlights the extent of apparent spatial disadvantage. Why are certain areas suffering more than others? It may be that a given region's problems are the result of the existence of a relatively high proportion of slow growing or structurally declining industries, compared with a relatively small expanding sector (this is typical of regions still dependent on heavy industries such as coal, steel and shipbuilding); it may be that a particular area's problems are the result of the actual movement of industry and the immobility of individuals whose jobs have disappeared (this is typical of many inner cities). Both cases illustrate why more disaggregate forward assessments are necessary to guide policy at the regional and local levels, albeit, where necessary, linked to national projections. The models outlined in this chapter attempt this by trying to explain why certain areas grow while others decline. However, they provide descriptions of outcomes and not explanations of causes. For this reason, a number of commentators have approached the issue rather differently by examining the changing structures of industrial ownership and investment decisions both nationally and internationally (see, for example, Massey and Meegan, 1982). While offering new insights into national employment change, this work has, as yet, limited applicability at the local level. It does, nonetheless, highlight the potential of investigations into corporate decision making in order to understand and forecast their local impact.

Employment change reflects changes in output, changes in productivity, changes in working practices, etc., but in no standard or easily predictable way. Loss of jobs may be the result of falling demand and thence output, or the result of mechanization, increased productivity, rising output and labour displacement. Moreover, the relationships between employment, invest- ment and production vary substantially from industry to industry, area to area, and over time. Translating such information into land requirements is particularly difficult without a more detailed understanding of the local economy. In this context, perhaps the most likely source of improvement in forecasting procedures may be through the further development, at a more local scale, of the kind of econometric forecasting techniques that have already proved so successful at the national level. Meanwhile, for the local authority planner, it is likely that employment forecasts will continue to be based on national forecasts, simple versions of the models outlined in this chapter, trend projections of individual industries, and separate considera- tion of industries or large employers of particular local importance.

5 Shopping

5.1 Introduction

Perhaps more so than with any other activity, land-use policies in respect of retailing can have a profound effect on profitability and investment. A new shopping development in one part of town might well flourish, but often at the expense of older established centres elsewhere. It is vital, therefore, that some coordination is exercised and, to this end, planners have developed increasingly elaborate techniques with which to approach retail planning. In assessing existing provision and making estimates of future needs and ways of meeting them, a variety of techniques are employed, but a common feature of most shopping studies is the use of gravity, potential and spatial interaction models. These models, which for short we shall term gravity models, lend themselves, in particular, to urban planning decisions that depend strongly on a knowledge of the relationship between activity locations and the travel behaviour of the users of such activities. They have not only been used by urban theorists to explain the structure and functioning of urban areas, but have also facilitated quantification thereby providing practising planners with a means of obtaining numerical results to specific problems of urban development.

5.2 The gravity concept of human interaction

5.2.1 The origin of gravity models

Gravity models are so-called because of an analogy with the ideas of Sir Isaac Newton whose Law of Universal Gravitation states that 'Two bodies in the universe attract each other in proportion to the product of their masses, and inversely to the square of their distance apart.' Mathematically, this relationship can be expressed as shown overleaf:

$$F = \frac{GM_1M_2}{d^2} \tag{5.1}$$

where F = the force of attraction between the two bodies
M_1 and M_2 = the respective masses of the two bodies
  d = the distance between the two bodies
  G = the gravitational constant.

Empirical studies made in the 1940s by Zipt (1949) and, independently, by Stewart (1947), showed that a number of human activities involving inter-action over space could be predicted, with surprising accuracy, by the laws used in Newtonian physics. However, in the application of the gravity concept to the analysis of urban systems, the gravitational force exercised by the two celestial bodies is interpreted as the amount of interaction between two areas, and the mass of the bodies is measured in terms of the respective areas' sizes or attractiveness. The proposition is then, quite simply, that the amount of interaction between two areas, A and B, is related directly to the size or attraction of the areas and inversely to the distance separating them. Mathematically, this relationship can be expressed as follows:

$$I_{AB} = \frac{K P_A P_B}{d_{AB}^{\lambda}} \tag{5.2}$$

where I_{AB} = the interaction between areas A and B
P_A and P_B = the respective sizes of areas A and B
      (measured, for example, in terms of population)
  d_{AB} = the distance between A and B
   λ = an exponent applied to the distance variable
   K = a constant which is empirically determined.

This formulation of the gravity principle has been applied to several aspects of the planner's work, but this chapter is concerned specifically with its application to retailing.

5.2.2 Reilly's law of retail gravitation
Long before Zipt and Stewart, William Reilly, a market research analyst, was employing a crude gravity formulation to delineate market areas around retail centres. Reilly was essentially concerned with the relative attractive-ness of two shopping centres to those people living between the centres. In its simplest terms he expressed his law as follows: '. . . under normal conditions two cities draw retail trade from a smaller intermediate city or town in direct proportion to some power of the population of these two larger cities and in an inverse proportion to some power of the distance of

each of the cities from the smaller intermediate city.' (Reilly, 1929, p. 16). The exponents used in connection with population or distance will obviously depend upon the peculiarities of any particular combination of retail circumstances, but from his empirical work Reilly found that typically: '. . . two cities draw trade from a smaller intermediate city or town approximately in direct proportion to the first power of the population of these two larger cities and in an inverse proportion to the square of the distance of each of the larger cities from the smaller intermediate city.' (Reilly, 1929, p. 16). Given the situation of choice faced by shoppers living between two cities, A and B, we can now say that the attraction of the shopping centre of city A, with population P_A, to individuals living at an intermediate location, distance d_A from A, will be

$$G_A = \frac{P_A}{d_A^2} \qquad (5.3)$$

where G_A = the force of attraction of A.

Similarly, the attraction of the shopping centre of city, B, with population P_B, to individuals living at an intermediate location, distance d_B from B, will be:

$$G_B = \frac{P_B}{d_B^2} \qquad (5.4)$$

where G_B = the force of attraction of B.

If an individual shops in the city with the strongest overall attraction, a simple formula can be used to find the market boundary or breaking point between these two retail centres. This is the point to which one city exercises the dominating retail influence and beyond which the other city dominates. It is the equilibrium point between A and B where the attraction of A is equal to the attraction of B, i.e. where:

$$G_A = G_B \qquad (5.5)$$

or, substituting (5.3) and (5.4) into the above expression, where:

$$\frac{P_A}{d_A^2} = \frac{P_B}{d_B^2} \qquad (5.6)$$

Shoppers living at this point will be indifferent between the two competing retail centres. Assuming that the market boundary is on a line connecting the

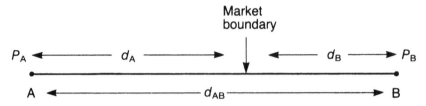

Figure 5.1   *A market boundary between two centres.*

two centres, as illustrated in Fig. 5.1, it can be calculated quite simply, using equation (5.6), as follows.

Given $\dfrac{P_A}{d_A^2} = \dfrac{P_B}{d_B^2}$ (5.6)

But $d_{AB} = d_A + d_B \Rightarrow d_A = d_{AB} - d_B$ (5.7)

Substituting for d_A in equation (5.6) we get

$$\frac{P_A}{(d_{AB} - d_B)^2} = \frac{P_B}{d_B^2} \Rightarrow \frac{P_A}{P_B} = \frac{(d_{AB} - d_B)^2}{d_B^2}$$ (5.8)

Taking the square root of both sides of (5.8) we get

$$\sqrt{\frac{P_A}{P_B}} = \frac{d_{AB}}{d_B} - 1$$ (5.9)

Adding one to both sides of (5.9) and cross-multiplying gives

$$d_B = \frac{d_{AB}}{\sqrt{(P_A/P_B)} + 1}$$ (5.10)

If the two cities A and B are 30 km apart and have populations of 100 000 and 25 000 respectively, we can substitute these values into our final expression, (5.10), to identify the market boundary, i.e.

$$d_B = \frac{d_{AB}}{\sqrt{(P_A/P_B)} + 1}$$

$$d_B = \frac{30}{\sqrt{(100\ 000/25\ 000)} + 1}$$ (5.10)

$$= 10$$

which means that B's market boundary lies 10 km away, from which it follows that A's market boundary is 20 km away. The solution is as illustrated in Fig. 5.2.

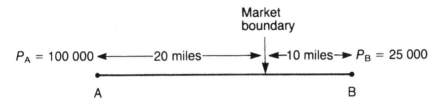

Figure 5.2 *Reilly's law applied to determine the market boundary.*

The problem with applying Reilly's law to delimit the trade hinterlands of retail centres is that it conveys the idea of discontinuous trading areas. The

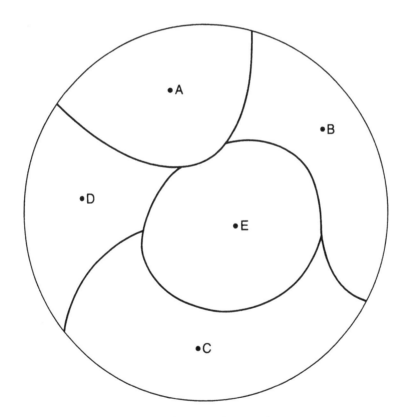

Figure 5.3 *Discontinuous trade areas for retail centres.*

model merely predicts a fixed number of shopping trips to a given retail centre and no trips to the centre beyond that boundary. This is illustrated in Fig. 5.3.

In practice, however, shoppers are not confronted with the simple bilateral choices that Reilly considered. The proposition that a line in space can be identified which separates population such that all those living to one side will make their purchases at one centre and all those on the other side at another, is fanciful. Even cursory observation demonstrates that trade areas for retail centres are not discontinuous, but rather that people have a choice of several competing centres, and that potential shoppers at particular locations do not always make the same choice — indeed, an individual shopper may even occasionally switch from one centre to another. But empirical evidence does show that the attractiveness of competing centres decays with distance and increases with scale of centre. For example, within a subregion containing a number of shopping centres, a given individual is more likely to visit a large centre where the quality, quantity and variety of shops best meets his or her needs, but might be discouraged from going to this centre if it is too far away. Despite the limitations of Reilly's approach then, these observations are consistent with his ideas. They imply that trading areas are, in fact, continuous and suggest the need for a probabilistic approach using the gravity model as a distribution model — a technique exemplified in the work of Casey (1955) and subsequently expanded by Huff (1963). In this way, attention is focussed on how consumers are attracted to a whole range of competing centres in an area, i.e. on the probability that they will visit any given centre within that particular subregion.

5.2.3 A probabilistic approach to the gravity model

A more general or probabilistic approach to the gravity model might take the following form:

$$T_{ij} = \frac{K P_i^\alpha H_j^\beta}{d_{ij}^\lambda} \tag{5.11}$$

This relationship simply says that, for a given time period, the number of trips (T) made by those living at i to an activity or centre located at j will increase with the number of trip makers (P) located at i, and with the number of opportunities (H) available at j to satisfy the demands of these trip makers. The number of trips decreases with the distance (d) or cost of travelling from i to j. In the model as specified, K is similar to our familiar gravitational constant, an α, β and λ are the other model parameters, all of which can be empirically determined in the calibration process. Methods of statistically estimating these parameters will be discussed later.

Readers will by now have noted that equation (5.11) above represents an extension of equation (5.2), although the latter was only concerned with the interaction between two centres. The more general form of equation (5.11) can simulate for any part of the subregion the interaction between activities, and is also capable of allocating activities to given localities or study zones. The variables used to serve as measures of P and H depend, of course, on the model's purpose — for example, on whether the interactions under consideration are recreational, educational or, perhaps, connected with the journey to work. Since this chapter is concerned with shopping, the interest is more likely to be with expenditure flows (S) than with trips (T). Because the value of purchases made by residents of i at shopping centre j is now being predicted, the money available for purchases (C) at i is inserted instead of (P), and the total retail floorspace (F) at j instead of (H). Retail floorspace is a common and reliable measure of attraction in shopping studies, but there is no reason why other factors, such as the availability of car parking, should not be incorporated in the index. The model may now be respecified as follows:

$$S_{ij} = \frac{K\, C_i^\alpha F_j^\beta}{d_{ij}^\lambda} \tag{5.12}$$

An important point to note at this stage is that in applying equation (5.11), an overall increase in the attractiveness (H) of location j has the effect of increasing the total number of trips made by the residents (P) of i. The model, or more specifically the potential value of T_{ij}, is *unconstrained*. But not all the activities which are considered when using gravity models have this effect. For example, the level of household expenditure does not depend on the locational pattern of shopping centres, but on other variables such as family income. In a retail distribution gravity model the overall level of money income, and hence the potential to spend within a subregion, is given, and we are concerned only with the distribution of this expenditure, i.e. with total sales in each shopping centre. These can be calculated from our model output, which is a matrix showing the flow of consumer expenditure from a number of residential zones (our is) to a range of shopping centres (our js). The model must, therefore, be *constrained* to ensure that the combined expenditure flows from the residential zones to the various shopping centres remain constant at the fixed level C_i (the expenditure available in each zone and hence the total expenditure available in the subregion). The constraining equation is:

$$C_i = \sum_j S_{ij} \tag{5.13}$$

Now returning to equation (5.12) and for simplicity assuming $\alpha = \beta = 1$,

$$S_{ij} = \frac{K\, C_i F_j}{d_{ij}^\lambda} \tag{5.14}$$

Substituting (5.14) for S_{ij} in equation (5.13) we can write

$$C_i = \sum_j \frac{K C_i F_i}{d_{ij}^\lambda} = K C_i \sum_j \frac{F_j}{d_{ij}^\lambda} \tag{5.15}$$

Cross multiplying we can obtain a value for K as follows:

$$K = \frac{C_i}{C_i \sum_j \frac{F_j}{d_{ij}^\lambda}} = \frac{1}{\sum_j \frac{F_j}{d_{ij}^\lambda}} \tag{5.16}$$

If we now substitute this constraint for K into equation (5.14), the final expression reduces to:

$$S_{ij} = \frac{C_i F_j/d_{ij}^\lambda}{\sum_j F_j/d_{ij}^\lambda} \tag{5.17}$$

As can be seen, equation (5.17) is a simple probabilistic expression which considers the effects of competition. The probability that people from a residential zone (i) will spend their money in a particular shopping centre (j) depends on the attractiveness of that shopping centre (F_j/d_{ij}^λ) compared with the overall attractiveness of all the shopping centres in the subregion ($\sum_j F_j/d_{ij}^\lambda$). The application of the constraint has ensured that the sum of all expenditure flows between any particular origin i and all possible destination zones j sums to the known demand generated at that origin, i.e. C_i. Readers should also note that the constraint implies that the spatial system is regarded as closed in that a closed set of j exists and ensures that no consumer expenditure leaves the subregion.

5.2.4 Applying the constrained gravity model

A fuller understanding of the application of the constrained gravity model in shopping studies is, perhaps, best gained by demonstrating its use with reference to a simple example.

Consider a hypothetical study area which, for the purpose of this analysis, has been divided into three zones. At present there are two shopping centres, A and B, located in zones 1 and 2 respectively, and a third centre C is proposed for zone 3. The situation is as illustrated in Fig. 5.4.

Planners are interested not only in the economic feasibility of the proposed centre but also in its impact on existing centres. Appropriate use of the model can assist with answers to both questions. To simplify the arithmetic it is assumed in this exercise that $\alpha = \beta = 1$ and $\lambda = 2$.

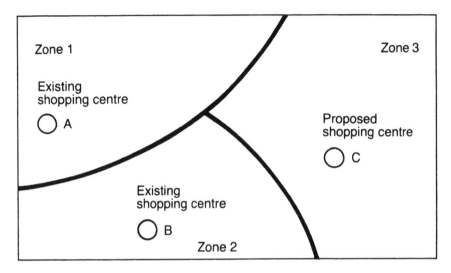

Figure 5.4 *The three-zone study area.*

Assuming that the proposed centre has been built and since the model deals with a closed system (boundaries are carefully defined so as to make the study area relatively self-contained), the residents of each zone will have just three shopping choices available to them. They can spend their money at the centre within their own zone or they can travel to the centres in either of the adjacent zones. The first stage in the analysis is to produce a *distance matrix* estimating the travelling distances involved in exercising each choice. Because the resident population within a zone is dispersed it is necessary to work with average distances and a zone centroid is defined (analogous to a zone's centre of gravity) from which it is assumed all trips originate. The distance to a particular shopping centre or destination can then be estimated directly. This is illustrated in Fig. 5.5, although readers should note that defining the zone centroid is not, in itself, a trivial exercise. The zones in actual shopping studies are unlikely to be neat geometric shapes and, furthermore, the population within these zones is unlikely to be evenly distributed. As a result, centroids often have to be estimated on the basis of informed guesses. The distance matrix is illustrated in Table 5.1 (overleaf) where, for example, it can be seen that residents of zone 1 must travel 2 miles to shop at A, 6 miles to shop at B, and 10 miles to shop at C. Although, in this example, the distances are straightforward, readers should once again note that, in actual studies, the deterrent effect of distance is often measured more accurately in terms of travelling time, or indeed, using the even more complex concept of *generalized cost* (see Section 5.3.3).

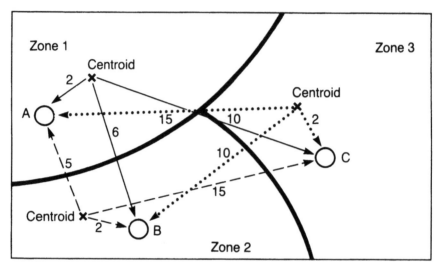

————— Shopping choices for residents of Zone 1
— — — Shopping choices for residents of Zone 2
········ Shopping choices for residents of Zone 3

Figure 5.5   *Travelling distances to shopping centres.*

Table 5.1   *The distance matrix (miles) for the study areas*

Zone \ Centre	Existing centre A	Existing centre B	Proposed centre C
Zone 1	2	6	10
Zone 2	5	2	15
Zone 3	15	10	2

The second stage in the analysis is to collect information on the relative attractiveness of the various shopping centres. As stated earlier, retail floorspace is often used as an index of shopping centre attraction although actual studies have occasionally elaborated on this simple index by making allowances for other factors including the availability of banks and other services, the presence of a department store, the availability of car parking and so on (see Section 5.3.3). Retail floorspace is invariably the dominant factor, however. Retail floorspace details for the shopping centres in our study area are listed in Table 5.2

Table 5.2   *Retail floorspace for shopping centres in the study area*

Shopping centre	Size of centre in ft² (m²)
A	10 000 (930)
B	90 000 (8400)
C	36 000 (3300)

The data compiled thus far can now be used to produce a matrix calculating the attraction measures F_j/d_{ij}^2 for each zone i and shopping centre j. For example, for zone 1 and shopping centre A the ratio is

$$\frac{F_A}{d_{1A}^2} = \frac{10\ 000}{2^2} = 2500$$

These attraction measures are listed in Table 5.3 where the total attraction that the three centres would have on each zone, $\sum_j F_j/d_{ij}^2$, has also been computed.

Table 5.3   *The F_j/d_{ij}^2 matrix for the study area*

Zone \ Centre	A	B	C	Total
1	2 500	2 500	360	5 360
2	400	22 500	160	23 060
3	44.5	900	9 000	9 944.5

The next stage in the analysis is to calculate the probabilities of residents in each zone shopping in each centre. Since the probability function is

$$\frac{F_j/d_{ij}^2}{\sum_j F_j/d_{ij}^2}$$

this is done simply by dividing the attraction measure of a given row and column of Table 5.3 by the total overall attraction measure found in the same row's total column. For example if, in the general form,

$$\Pr \begin{bmatrix} \text{a resident of zone } i \\ \text{shopping at centre } j \end{bmatrix} = \frac{F_j/d_{ij}^2}{\sum_j F_j/d_{ij}^2}$$

then, for this particular exercise:

$$\Pr \begin{bmatrix} \text{a resident of zone 1} \\ \text{shopping at centre A} \end{bmatrix} = \frac{F_A/d_{1A}^2}{F_A/d_{1A}^2 + F_B/d_{1B}^2 + F_C/d_{1C}^2}$$

$$= \frac{2500}{2500 + 2500 + 360}$$

$$= 0.4664$$

These probabilities, which measure the relative attractiveness of the various centres, are listed in Table 5.4.

Table 5.4　*The shopping probability matrix $F_j/d_{ij}^2 \big/ \sum_j F_j/d_{ij}^2$ for the study area*

Zone \ Centre	A	B	C
1	0.4664	0.4664	0.0672
2	0.0174	0.9757	0.0069
3	0.0045	0.0905	0.9050

To complete the analysis, further data is required on the availability of retail expenditure within each residential zone. This is usually compiled by multiplying the population or household population in each zone by an average figure for consumer retail expenditure per head or household, as appropriate. Data on existing population can be obtained from the Population Census although, for projective exercises, a population forecast, using one of the techniques discussed in Chapter 2, might be required. If the analysis is based on households, these population figures will obviously have to be adjusted. Crude average expenditure figures can usually be calculated using data from the Family Expenditure Survey although, in projective models, these have to be forecast as before, usually by applying a compound growth rate (a version of a trend line) to base-year data. Table 5.5 sets out the consumer expenditure, C_i, available for each of the three zones in our study area.

Table 5.5 *Consumer expenditure, C, for the zones in the study area*

Zone	Population	Average annual retail expenditure per head (£)	Total consumer retail expenditure C_i
1	400	100	40 000
2	300	100	30 000
3	300	100	30 000

In exercising their shopping choices, the residents of each zone will distribute their expenditure between centres in the proportions previously calculated in the probability matrix. If the values in each row of the probability matrix are multiplied by the consumer expenditure available in each residential zone, this will give a new S_{ij} matrix, where

$$S_{ij} = \frac{C_i \, F_j/d_{ij}^2}{\sum_j F_j/d_{ij}^2}$$

showing the predicted flow of retail expenditure from each zone i to each zone j. For example, for the flow of expenditure from zone 1 to centre A:

$$S_{1A} = \frac{C_1 F_A/d_{1A}^2}{F_A/d_{1A}^2 + F_B/d_{1B}^2 + F_C/d_{1C}^2}$$

$$= (40\ 000)\ (0.4664)$$

$$= 18\ 656$$

The results are summarized in Table 5.6 (overleaf), which is the final output from the retail distribution gravity model — a matrix showing the flow of consumer expenditure from each residential zone to every shopping centre and the total sales in each centre.

The planner is now equipped to deal with a whole range of policy questions. One is clearly in a position to evaluate the new centre by comparing the predicted turnover with any minimum threshold that has been defined as necessary to ensure its viability. Furthermore, the impact of the new centre on existing centres can be judged by comparing the model's output with the actual or predicted expenditure flows that would have occurred in the absence of the new development. The overall process is summarized in the flowchart reproduced as Fig. 5.6 (overleaf).

Table 5.6   *The S_{ij} matrix, expenditure flows and total sales for the study area*

Zone \ Centre	A	B	C	Total consumer retail expenditure (£)
1	18 656	18 656	2 688	40 000
2	522	29 271	207	30 000
3	135	2 715	27 150	30 000
Total retail sales (£)	19 313	50 642	30 045	100 000

5.3 Empirical realities

5.3.1 Parameter estimation and the problem of calibration

Thus far, the problems of parameter estimation have been avoided since it has been assumed throughout that $\alpha = \beta = 1$ and $\lambda = 2$. However, variations in the sizes of these parameters can have a profound impact on a model's output and the generalized model obviously needs to be calibrated to fit the real world and the peculiarities of the place and time under consideration. This process of calibration is often achieved by the trial and error process of adjusting the model's parameters to find the best fit between a series of model outputs and an actual base-year situation. Returning to the original generalized model, equation (5.11), the procedures can be described more scientifically as follows (less numerate readers may ignore the following paragraphs and proceed directly to Section 5.3.2).

$$\text{Given} \quad T_{ij} = \frac{KP_i^\alpha H_j^\beta}{d_{ij}^\lambda} = KP_i^\alpha H_j^\beta d_{ij}^{-\lambda} \tag{5.18}$$

Taking logarithms of both sides of this equation we get

$$\log T_{ij} = \log K + \alpha \log P_i + \beta \log H_j - \lambda \log d_{ij} \tag{5.19}$$

By performing this straightforward logarithmic transformation, what at first appeared to be a rather complex equation is converted into a simple linear relationship in the logs of the variables T_{ij}, P_i, H_j and d_{ij}. Assuming time series or cross-section data are available on these variables, a multiple regression of the form:

$$\log T_{ij} = a + b_1 \log P_i + b_2 \log H_j + b_3 \log d_{ij}$$

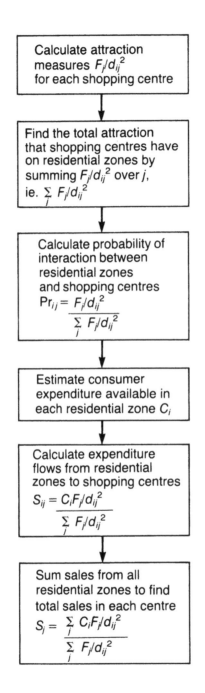

Figure 5.6   *Flow chart for typical shopping model.*

will yield values of the regression coefficients in which b_1 is an unbiassed estimate of α, b_2 an unbiassed estimate of β and b_3 an unbiassed estimate of $-\lambda$. Moreover, the regression process will enable us to say something about the quality of these estimates since it will also produce t values of the coefficients and the appropriate measures of correlation. However, there is a problem in using the constant of regression a to derive an estimate of K. As Krueckeberg and Silvers (1974) point out, the antilog of a yields an over-estimate of K, a problem that can easily be overcome by employing their methodology. Denoting the observed values of the variables $\hat{T}_{ij}$, $\hat{P}_i$, $\hat{H}_j$ and $\hat{d}_{ij}$ and using the unbiassed estimates of α, β and λ from the regression, they proceed as follows. If K were like the other parameters and unbiassed, the model as described in equation (5.18) would yield unbiassed estimates T_{ij}^*. The expected values of the estimates obtained from the model $E\ (T_{ij}^*)$ must, therefore, be equal to the average value of the observed $\hat{T}_{ij}$. Using the observed values $\hat{P}_i$, $\hat{H}_j$ and $\hat{d}_{ij}$ in equation (5.18), summing over n observations $(j = 1, \ldots, n)$ and dividing by n gives

$$E(T_{ij}^*) = \frac{\sum\limits_{j}^{n}\hat{T}_{ij}}{n} \qquad \frac{K\sum\limits_{j}^{n}\hat{P}_i^\alpha\hat{H}_j^\beta\hat{d}_{ij}^{-\lambda}}{n} \tag{5.20}$$

Cross multiplying and cancelling the ns we get:

$$K = \frac{\sum\limits_{j}\hat{T}_{ij}}{\sum\limits_{j}\hat{P}_i^\alpha\hat{H}_j^\beta\hat{d}_{ij}^{-\lambda}} \tag{5.21}$$

which is an unbiassed estimate of K.

The subject of parameter estimation is a very technical and difficult one, and a more detailed discussion is beyond the scope of this text. It is, nevertheless, useful to note what values have been obtained in practice, and Table 5.7 includes examples from the more popularly quoted studies. The values are derived from the NEDO (1970) report which, although now somewhat dated, is often cited, as it is in many respects the seminal work in this area.

5.3.2 The problem of small distances
An important problem often arises with the distance attenuator and the application of the inverse power law. For example, with our standard model, when d approaches zero the value of $1/d^\lambda$ gets larger, and therefore the prediction of T_{ij} becomes larger and larger (approaching infinity). More precisely, the inverse function $1/d^\lambda$ underestimates the deterrent effect of small distances. This situation is illustrated in Fig. 5.7(a). To overcome the problem of small distances, it is necessary to choose a function which is defined when d is equal to zero and analysts have frequently suggested using

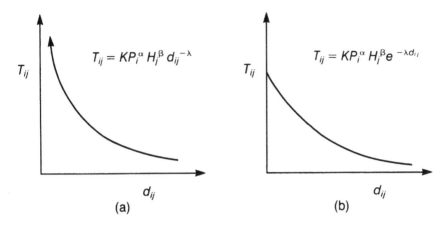

Figure 5.7 *The effect of changing the parameter of the distance function.*

the negative exponential function. For example, Black (1966) developed a submodel of a more general model predicting land use and traffic flows which was essentially the same as the previously described gravity models but postulated a deterrence function having an exponential form.

But what is a negative exponential? It is called a negative exponential because the exponent is negative and it takes the form a^{-d} where a is a constant. So in our standard model, instead of writing $d_{ij}^{-\lambda}$, we would write $e^{-\lambda d_{ij}}$ where e is the constant 2.718 . . . It is immediately apparent that, when distance is equal to zero, the negative exponential is still defined — it always has a postive value. In other words, the function predicts a finite zero distance value of T_{ij} as shown in Fig. 5.7(b).

Using the negative exponential the standard model can be written as:

$$T_{ij} = K \, P_i^\alpha H_j^\beta \, e^{-\lambda d_{ij}} \tag{5.22}$$

Once again a straightforward logarithmic transformation can be performed which, using the base e, gives:

$$\log_e T_{ij} = \log_e K + \alpha \log_e P_i + \beta \log_e H_j - \lambda d_{ij} \tag{5.23}$$

Readers will have noted that the distance variable is not transformed and the regression now becomes:

$$\log_e T_{ij} = a + b_1 \log_e P_i + b_2 \log_e H_j + b_3 d_{ij}$$

which can be estimated as before.

5.3.3 Operational models and the definition of variables

When using a gravity model, definition of the attractiveness (mass) and deterrence (distance) variables will depend on the model's purpose. But even when this has been determined there are often a number of ways in which these variables can be specified. Indeed, definitions of this kind have been a major concern of many of the operational models used in shopping studies.

In operational shopping models, analysts have tended to use three particular measures as proxy variables for attractiveness: floorspace, sales, or some index of the type/mix/number of shops. Floorspace figures are relatively easy to obtain and it should come as no surprise that this measure is particularly favoured. Several publications contain official statistics on existing shopping floorspace but, perhaps more importantly, in projective models floorspace figures can be obtained directly from the development proposals set out in any planning applications. However, floorspace does suffer from certain disadvantages as a measure of shopping centre attractiveness, not least because its intensity of use and quality may vary considerably both within and between centres. Furthermore, the NEDO publication on shopping models pointed out an even more fundamental problem with floorspace as a proxy, namely its sensitivity over time.

> For example, the floorspace provision may remain constant through a period of declining sales due to population redistribution, and a subsequent reduction in attractiveness. (NEDO, 1970, p. 52).

Sales volume is at once a more precise measure but also a more difficult one in projective exercises. There are prediction difficulties because of a tautological problem that arises since sales are, in any case, a product of the model's output. The problem is highlighted by the NEDO where it is suggested that the use of sales volume

> . . . is subject to difficulties in prediction since, once introduced, it tends to logically imply a solution in which the result corresponds as closely as possible with the value assumed for attractiveness. (NEDO, 1970, p. 52).

The third and final measure acknowledges the catalystic effect of certain types of shop or service on a centre's attractiveness — for example, department stores, large variety stores, availability of banks, car parking, etc. — and attempts to embrace these factors in a single index. The problem with this approach is that with projective models the mix of shops is rarely predetermined and the index is accordingly difficult to specify.

In selecting appropriate proxy variables for deterrence, analysts have tended to use either the simple measure of straight line distance or a more complex measure of journey time by various forms of transport. Obviously physical distance alone ignores a number of other important considerations which might deter shoppers in metropolitan areas: e.g. factors such as congestion, physical barriers, the availability of public transport, car owner-ship etc., will all have an impact. Hence the tendency to use journey time. Indeed, in recent years, analysts have favoured the even more complex measures of economic distance or generalized cost (as discussed in Chapter 6) — i.e. what a traveller perceives it costs him or her to travel from A to B. For example, in the case of a car trip it would comprise: driving time cost, operating cost, parking charges, cost of extra time spent walking from car parks to actual destinations, and so on.

As with the measurement of attractiveness, the choice of proxy variable for the measurement of deterrence depends on the problem under consider-ation, the availability of data, and other related matters. However, there is considerable justification for using the simple measure, not least because of the close correlation found in empirical tests between journey time, general-ized cost, and straight distance.

Table 5.7 includes examples of the variables used in some of the more important theoretical studies and operational models.

Table 5.7   *Operational shopping models*

Model	Variables		Parameter estimates	
	Attractiveness	*Deterrence*	β	λ
Reilly model	Population	$(\text{distance})^{-2}$		
Huff model	Floorspace	$(\text{journey time})^{-2}$		
Haydock study	Sales	$(\text{journey time})^{-2}$		
Haydock model (durables trade)	(combination of indices)$^{\beta}$	$(\text{journey time})^{-\lambda}$	3.0	2.6
Lewisham model (all trade)	Sales	$(\text{distances})^{-\lambda}$		1.1
South Bedfordshire model (all trade)	Floorspace	$(\text{journey time})^{-\lambda}$		1.3
Teeside model (durables trade)	$(\text{Floorspace})^{\beta}$	$(\text{distance})^{-\lambda}$	1.38	2.36
Black model	$(\text{Sales})^{\beta}$	$e^{-\lambda \, (\text{distance})}$	0.95	0.20

5.4 Limitations of shopping models

In recent years there has been a considerable development in the degree of sophistication with which analysts approach urban modelling, and the plan-ning of retail provision is no exception. There seems to be a belief that with more and better quality information it is possible to design increasingly realistic models and calibrate them more finely for purposes of prediction. However, as David Thorpe has suggested '. . . the apparent sophistication of models often makes us forget how tenuously they represent the processes of the real world, even at their base point — let alone at some future date.' (Thorpe, 1975, p. 43). This is particularly true of shopping models and, in any critique of retail forecasting techniques, two aspects merit consideration. First, it is necessary briefly to address a number of the theoretical limitations of the gravity concept. Secondly, there is a consideration of some further important operational problems of shopping models not yet discussed.

5.4.1 Theoretical limitations

As has already been pointed out, the gravity and potential concepts of human interaction were originally developed from analogy to the Newtonian physics of matter. The development of this approach to so-called *social physics* was traced to the work of Reilly and others who identified a whole range of empirical regularities which formed the basis for the kind of equation used throughout this chapter. It can be argued, though, that the models only represent an empirical regularity to which it has proved difficult to furnish any theoretical explanation. In other words, gravity models describe the spatial interaction of activities within the urban economy rather than explain the behavioural reasons for any given pattern of location. Although they may be useful when used as a descriptive tool, therefore, gravity models give rise to fundamental problems when used in prediction. For example, in the absence of any plausible theory, how can the values or functions which are assigned to the exponents in the calibration process be explained. As Walter Isard has said

> . . . the justification for the gravity model is simply that, everything else being equal, the interaction between any two populations can be expected to be directly related to their size; and since distance involves friction, inconvenience, and cost, such interaction can be expected to be inversely related to distance. (Isard, 1960, p. 515)

Wilson (1970) has attempted to address the gravity model's lack of a theoretical basis by employing the entropy-maximizing principles of statistical mechanics. The physical law of entropy provides a macro-analytical description of the movement of gas particles under certain conditions. Clearly there are many possible distributions that satisfy any given set of conditions, and entropy-maximization determines the most probable. Wilson likens the movement of such gas particles to the movement of individuals and, in a shopping study, the method deals with these individuals and assesses their probability of making a particular shopping trip as a statistical average. In this way, the approach gives a statistical explanation and justification for spatial interaction by changing the basis of the simple

Newtonian analogy based on zonal aggregates (masses), and dealing directly with the actual components of the system of interest. The method can be used to derive statistically the whole family of gravity models which are, indeed, now generally referred to as entropy-maximizing models. However, notwithstanding these developments and the other advances that have been made in spatial interaction and model calibration, gravity models still only provide a statistical explanation rather than a behavioural reason for the spatial interaction of activities within the urban system.

Another criticism of gravity models, as applied in shopping studies, relates to their partial nature. The method necessarily simulates interaction of one activity, which is modelled, while other parts of the urban system are held constant. Obviously a more general approach would be preferable — that is an approach which explicitly acknowledges the extent to which several components/activities interact simultaneously. The Gavin–Lowry model and its fraternity represent attempts to achieve this more general approach, but in essence even these techniques are still essentially partial in nature.

A number of commentators have criticized the static nature of gravity models, arguing that they fail adequately to embrace the dynamic elements of the urban economy. According to this view, mathematical consistency is not enough: models need to embrace temporal and cause-and-effect modes of explanation. This kind of critique can obviously be applied to several of the models described in this text (see, for example, the discussion of economic base theory in Chapter 4) and is addressed more fully in Sayer (1976).

A further theoretical criticism focusses on the level of aggregation. It is obvious from our derivation that the gravity model is designed to consider activities at an aggregate level dealing, in this case, with population, expenditure and retail activity. This tends to obscure the idiosyncrasies of individuals and small groups which can be evaluated only if disaggregated models are developed. Unfortunately disaggregation creates problems as the resultant models are more complex and require considerably more data.

It is worth reiterating that since retail gravity models are singly constrained, they are essentially demand models in that activities are allocated to destination zones without reference to the level of supply in these zones. Further constraints can be applied, however, by restricting the amount of activity allowed to locate in particular destination zones as in the case of the trip distribution model described in Chapter 6. The models also deal with a closed region, that is all interaction takes place within the study area with no trips crossing the study boundary into the wider hinterland. This condition can be relaxed through the introduction of external zones but, once again, this imposes additional data requirements and further increases the complexity of the model formulation. Finally, it should be pointed out that the definition of study zones is itself not a trivial problem, since accurate definition can have a profound impact on model performance. For example, small zones make it easier to locate accurately zone centroids, whereas large zones facilitate the measurement of zonal variables where accuracy depends on the ability of zonal averages to satisfactorily submerge the importance of decision making by small groups or individuals.

5.4.2 Practical and operational limitations

A number of the practical or operational problems of gravity models have already been discussed: for example, the definition of variables, the difficulty of calibration and so on. But in shopping studies, two further issues give rise for concern. First and foremost there is the problem of determining levels of consumer expenditure. The latter depend crucially on accurate forecasts of consumer-disposable income, which are not only subject to wide regional variation, but are also very difficult to estimate if the economic climate is unsettled. Even if reliable estimates of disposable income are available, the relationship between income and expenditure is far from stable and is likely to fluctuate quite markedly given the current pace of social and economic change. Clearly the straightforward extrapolation of past trends and the application of simple averages can lead to very mistaken conclusions.

On this question of consumer expenditure, David Thorpe raises two further issues:

(a)   Does the quality of shops affect expenditure levels?
(b)   Does the quality of shops affect inflation?

(Thorpe, 1975, p. 45).

Intuitively a positive answer might be given to the first question, and empirical evidence tends to support this. Yet few forecasts allow for this impact on expenditure levels. As far as Thorpe's second question is concerned, the relationship between inflation and retail provision is more complex. But it does raise an interesting paradox:

> ... if prices are held down this could suggest that consumer purchasing falls in such an area — so less floorspace is apparently required simply because of lower prices! In reality, of course, lower prices release more income for expenditure and so rather than one item two may be bought and more floorspace is required. (Thorpe, 1975, p. 45)

Although it is difficult to make appropriate allowance in this case the application of past averages can, once again, lead to misleading results.

In the model developed in Section 5.2 expenditure flows and hence turnover levels were calculated but, in general, the final output of shopping studies is needed by planners in terms of floorspace, not sales. It is often necessary, therefore, to convert turnover figures into the equivalent amount of floorspace through the application of floorspace conversion factors (sometimes referred to as sales densities). For this purpose, most shopping studies distinguish between *convenience* and *durable* goods and average sales densities (normally measured in £ per square foot per annum) are calculated for each category from published data. Knowledge of existing and expected changes in these sales densities is invariably the most sensitive of the statistical assumptions made in shopping studies and is the second of the important operational limitations. Two problems arise. First, in judging what level of sales density is right at present, there is the perennial problem of using averages when empirical evidence suggests there are considerable differences between retailers in floorspace efficiency, that is sales per square foot. Secondly, in estimating future sales densities for predictive exercises one needs to adjust current figures. Most analysts support an annual increase in floorspace efficiency of 1% for convenience trades, 1.5% for durable trades, and 1.3% for all shops. Unfortunately, recent empirical evidence suggests extremely wide variations in floorspace efficiency and, indeed, even such a supposedly successful retailer as Sainsbury's has, in recent years, shown a decline in floorspace usage. It would appear that this decline is simply the result of opening new and larger stores, which have a lower turnover per square foot. However, it does, once again, demonstrate the misleading nature of averages. In the words of Colin Arnott and James Williams

> ... average floorspace conversion factors ... should be treated with extreme caution. They conceal wide variations of performance by each retailer and are of little or no assistance in assessing the impact of the individual proposals. (Arnott and Williams, 1977, p. 172)

5.5 Conclusions

Despite the obvious problems of developing and using spatial interaction models in shopping studies they are widely used and with good reason. They can be made very simple and easy to work with, can be calibrated quite accurately, and provide useful and broad quantitative answers to policy makers in determining future strategy. However, the use and application of such techniques must be seen in an explicit policy context. Retailing has undergone dramatic restructuring during the last decade and these changes have important implications for shopping provision. The techniques described in this chapter are demand models and, although useful aids in assessing floorspace needs, they completely ignore what could be called supply-side considerations, that is, they say little about the actual economics of retailing. They should, therefore, be supplemented by qualitative appraisals of shopping facilities to identify the commercial and other factors influencing the shopping structure. The problem is one of determining and reconciling consumer demands and needs with commercial trends.

6 Transport

6.1 Introduction

One of the most important characteristics of transport is that it is not usually demanded in its own right. Few people travel simply to enjoy the journey. Trips are normally made to take advantage of opportunities that are available at particular destinations be they social, recreational, educational or commercial. In other words, the demand for transport is derived because it reduces the spatial disadvantage of separation by improving the potential for communication between land-use activities. It is this intimate connection between land use and transport that makes the control or expansion of transport facilities one of the most powerful geographically specific instruments that planners can use to guide urban development. However, the dynamics of land use and transport interaction is very complex and it is not always possible to establish cause-and-effect relationships. For example, the effects of improvements in transport infrastructure obviously have a profound impact on land-use activities and land values, but the converse is also true — major changes that occur in land use, by either altering the size and distribution of the residential population or shifting industrial locations, etc., tend to change the pattern of transport demand. The primary objective of transport planning must, therefore, be to ensure that there is an efficient balance between land-use activities and the potential for communication between such activities.

6.2 Land-use transport studies

6.2.1 An overview of the modelling process

The medium of interaction between land-use activities and transport facilities is traffic and the objective of a transport study is, quite simply, to predict future traffic levels and to plan facilities to accommodate this traffic.

It is especially necessary to plan well and provide the right network capacity, since investment in transport infrastructure projects is often extremely long lived. To this end, it would be ideal to capture in a single analytical model the dynamics of land use and transport interaction, but no such model exists. Transport studies, therefore, attempt to describe and predict travel patterns using a series of linked submodels which simulate the decision-making process which an individual traveller might be expected to use when considering making a journey. Although the basic form of these studies is now fairly well established, the details of procedures and techniques used in particular cases can vary considerably depending on the peculiarities of the problem under consideration.

In general, it can be said that traffic forecasting attempts to predict the volume of traffic T from some point i, to another point j, by a given mode (be it public or private transport) k, on a particular route l. A typical study breaks down this aggregate term T_{ijkl} into its component parts and then attempts to deal with each component in a sequential traffic forecasting scheme. Generation, distribution, modal split and assignment can all, therefore, be considered as linked submodels together making up the transport study as illustrated in Fig. 6.1 (overleaf). This description is necessarily an oversimplification. In practice the stages are not as clear cut and discrete and, indeed, need not follow the sequence illustrated. For example, modal split is often handled in conjunction with one of the other three submodels, such as separately generating trips by mode or separately distributing them by mode.

Despite their interdependence in the process as a whole, the various stages are nevertheless best considered individually. However, before proceeding with a more detailed consideration of the overall study method, under the submodel headings outlined above, it is necessary to address the question of study design and basic data requirements. The basis for a systematic approach in any transport study is provided in two ways: first, by very careful study design, including physically subdividing the whole study area into zones and treating movement from one zone to another as a trip; and, secondly, by the collection of data via various surveys including, especially, detailed field observations of what trips are actually made, by whom, for what purpose, and by what means.

6.2.2 Study design
Three important issues need to be considered under this heading: the definition of the study area and identification of boundaries; the division of the study area into zones; and the specification of road and public transport networks.

(a) Study area
As with most of the problems encountered in this text, administrative

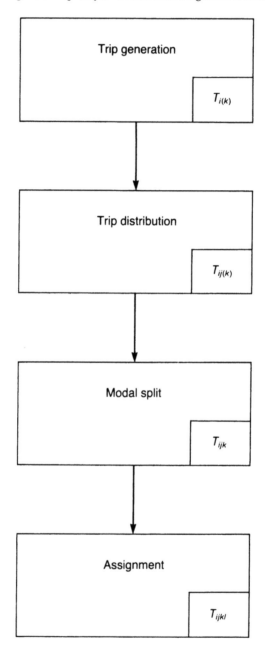

Figure 6.1 *The traffic-forecasting process.*

boundaries, while undoubtedly convenient, are not always the most relevant for purpose of a study. Although the actual definition of an appropriate boundary is largely a matter of subjective judgment, there are, nevertheless, some general rules which should be followed. First, the study area should be as self-contained as possible, that is, it should be defined so as to minimize interaction across the boundary (also known as the *external cordon*). For example, in an urban transport study, the boundary should be drawn fairly tight around the urban area but should, where possible, anticipate and embrace future growth areas. The determination of the boundary should also take advantage of man-made and natural perimeters which limit cross-boundary movement: for example, rivers, railways, motorways, etc. Finally, the inclusion of peninsulas and inlets should be avoided. As Wells notes:

> Peninsulas might be completely crossed by external traffic, which would thus be observed twice yet leave virtually no effect on the internal area require-ments. Inlets might well be crossed twice by internal traffic and at the same time, themselves form peninsulas nearby. (Wells, 1975, p. 70)

In any event, in both cases there will be confusion in balancing traffic flows as illustrated in Fig. 6.2 (overleaf).

(b) Zoning

Once defined, the study area then has to be divided into zones. Because of the obvious difficulty of handling and analysing survey data by individual property or address, a zoning system is a necessity. But, as pointed out in previous chapters, zoning is not a trivial exercise.

Probably the most important factor is that zones should be as homogen-eous as possible, since data will be used to describe their average land-use characteristics. This may mean some zones having odd shapes to facilitate the assembly of planning data and to ensure the preponderance of a single planning use. Zonal size is also very important. If zones are too large, most trips will take place within rather than between zones and intrazonal trips cannot be readily modelled. However, if zones are too small, not only do the interaction flows between zones become so complex that aggregate models have difficulty in describing trip patterns, but the statistical significance of the data also becomes questionable. Related very closely to this question of zonal size is the actual number of zones. Clearly, too many zones will considerably increase the complexity of the modelling process and make it more difficult to interpret and understand. On the other hand, too few zones, by increasing the proportion of zones on the edge of the study area, also increases the probability of interaction across the boundary and threatens 'self-containment'. That is not to suggest that the modelling process does not include a consideration of the wider area. Indeed, most studies explicitly acknowledge this and also include a series of external zones. As a general

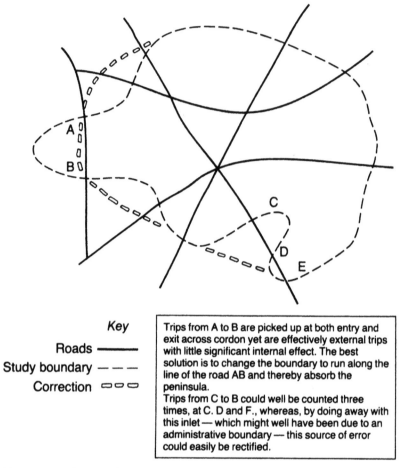

Key

Roads —————

Study boundary — — —

Correction ▭▭▭

Trips from A to B are picked up at both entry and exit across cordon yet are effectively external trips with little significant internal effect. The best solution is to change the boundary to run along the line of the road AB and thereby absorb the peninsula.

Trips from C to B could well be counted three times, at C. D and F., whereas, by doing away with this inlet — which might well have been due to an administrative boundary — this source of error could easily be rectified.

Figure 6.2   *Peninsula and inlet effects on study boundaries. (Reproduced by permission of the publishers, Charles Griffin & Co. Ltd, 16 Pembridge Road, London, from Wells, G. R.:* Comprehensive Transport Planning *(1975).)*

rule, in transport studies *internal zones* are usually smallest nearest the centre of the study area, increasing in size towards the boundary. *External zones* are often subdivided into intermediate and external zones — a recognition of the fact that one needs a finer zoning system adjacent to the study area (*intermediate zones*) than the truly remote external zones in the rest of the country. Finally, zones should have regard to major routes as well as other natural and administrative boundaries.

(c)   Network definition

Defining the study area and splitting it into zones produces an acceptable representation of the land-use system that is amenable to further analysis. It is now necessary to perform a similar exercise with the road and public transport networks, which also need to be defined, disaggregated and codified, so that the land-use and transport systems may be brought together to provide the essential building blocks for the modelling exercise.

As regards the road network, it is not possible to include every road within the study area. Only those which are considered significant to traffic are, therefore, selected. The selected roads are then divided into *links* and *nodes*. Nodes are provided at all important intersections in the network and also at points along particular routes where the character of the road might change, for example, where a single carriageway road becomes a dual carriageway. Nodes are then numbered serially, enabling network links to be identified by the nodes at each end, and also facilitating the specification of the connectivity properties of the network in a node–node matrix (as illustrated, for example, in Fig. 6.7 and Table 6.8).

The public transport network is very different from the road network, scheduled stopping points and route interchanges being more important than road junctions and links. For example, although subject to the constraints of the existing road network, buses suffer a further limitation in that they have to operate on predetermined routes. Furthermore, passengers can only join, leave, or change directions at stopping and interchange points.

Obviously trip makers will need to join the transport networks that link land-use activities when travelling between and within zones. Each of these network-joining trip parts cannot be handled separately because of the resultant increase in the complexity of the modelling process. For this reason it is assumed that all trips generated within a zone, or attracted to it, emanate or terminate from the zone centroid (a similar assumption was made when developing the shopping model in Chapter 5), which is linked by a *centroid connector* to a node or nodes as appropriate.

The study area and transport system has now been specified and codified in a way that lends itself more readily to analysis and synthesis, and the collection of basic data can proceed.

6.2.3 Collection of basic data

As stated earlier, the demand for transport is a derived demand. Traffic levels can, therefore, be thought of as a function of land uses and buildings, and it is this factor which is the basis of transport planning. The transport study process simply attempts to establish quantifiable relationships between land use and travel demand by calculating present-day trip generation rates for different uses and then extrapolates such relationships to produce estimates of future travel demands, together with possible ways of meeting them. A number of surveys need to be undertaken, therefore, to assess the present demand for movement and how it is met, and so identify what relationships exist between the characteristics of the study area and trip making *per se*. Four categories of information are required: existing travel patterns, household and population characteristics, an inventory of existing transport facilities, and land-use data.

(a)   Existing travel patterns
Travel pattern data is required for four basic movements:

1   *internal–internal*, that is between one internal zone and another;
2   *external–internal*, that is between one external zone and an internal zone;
3   *external–external*, that is between one external zone and another;
4   *intrazonal*, that is within individual zones.

All of these movements can be made by different modes of transport and, depending on the nature and complexity of the particular transport study, it may be necessary to collect data on each type of movement for each mode. In any event, the data is collected via a range of surveys, including roadside interviews and associated traffic counts, household interviews, commercial vehicle surveys, and public transport surveys.

(b)   Household and population characteristics
The Population Census and other published statistics can provide an analyst with a certain amount of data, but for more detailed information covering basic household characteristics and travel habits, a *household interview* (HI) is usually necessary. Not every household in the study area need be interviewed, and appropriate sample surveys are normally undertaken. The household interview is designed to ascertain details of the size and structure of households, the occupation and employment structure of household occupants (including place of work), the place of school/further education for those of school age, car ownership levels, household incomes and, most importantly, a record of all journeys made on a given day by all members of the household over 5 years of age (specifying mode).

(c)   Existing transport facilities
Clearly a knowledge of existing transport facilities is essential. The third category of basic data is, therefore, a stocktaking exercise of major highway and public transport networks.

As far as the inventory of highway facilities is concerned, information is needed on link lengths and carriageway widths, speeds and capacities, accident rates, traffic regulation and junction spacings, the nature and density of adjacent development, and the frequency of frontage access. Much of this information can be obtained by desk study, but a traffic volume census and travel-time surveys are also likely to be necessary.

Much of the information required for the public transport inventory can be obtained from published route maps and timetables. But accurate estimates of the capacities of various services can be obtained only by field studies and/or the cooperation of the operators in question.

A final and most important stocktaking exercise under this heading is a parking survey. The availability of parking facilities can have a marked influence on travel demand, and information needs to be collected not only on physical location, but also on the type, capacity and operating characteristics of existing on- and off-street facilities.

(d)   Land use and other planning data
Having compiled most of the travel data required, it is now necessary to collect the corresponding planning and economic data.

Since the amount and characteristics of travel are influenced by the nature and intensity of the development of land, the first requirement is to compile basic land-use information. For residential areas, the intensity of use is normally measured in terms of net residential density, whereas in employment zones and shopping areas the measurement tends to be in terms of plot ratio or an appropriate floorspace index. Most of the data required is readily available in official publications and can be updated with reference to the statutory register of planning applications when necessary.

For each internal traffic zone, basic population statistics are needed, including an indication of household composition. Information as to the size and structure of the labour force is also required, as are other details already specified above. Much of this information can be derived from the Population Census (including the Sample Census), but because of the incompatibility of enumeration of district boundaries and traffic zones, most of the information is usually obtained in a household interview survey.

6.2.4 The transport planning process
In summary, the study process can be described as in Fig. 6.3 (overleaf) which is an elaboration of the sequential forecasting scheme illustrated in Fig. 6.1 with the addition of the study design and data collection stages.

6.3 Traffic forecasting

With codification of the study area and basic data collection complete, traffic forecasting can proceed and this will be considered in more detail under the main submodel headings.

6.3.1 Trip generation
The first of the submodels in the conventional study process is that which predicts the number of trips starting and finishing in each zone, i.e. the value of T_i in Fig. 6.1. But before looking at trip generation techniques in more detail, several terms need to be defined — what is a trip, what makes trips, and when do trips take place?

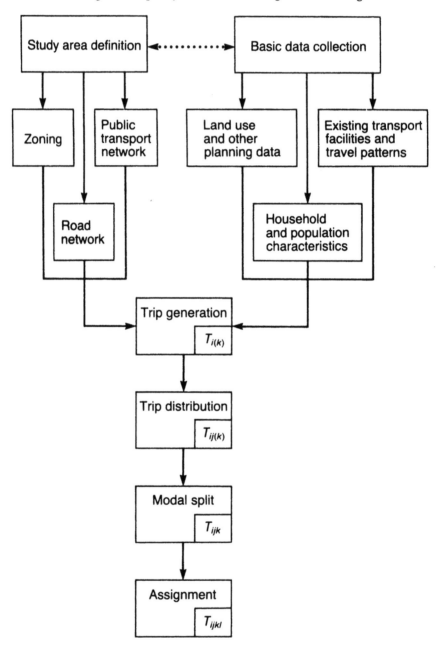

Figure 6.3 *The transport study process.*

A trip can be defined in a number of ways. It is not uncommon to hear people speak of a round trip, e.g. to go shopping and then return home. One could define a trip as movement between an origin and a destination, in which case the shopping visit would contribute two trips, i.e. the initial trip to the shops followed by the return journey. However, as part of the trip generation process it is normal practice to work in trip-ends (every trip, of course, has two ends) and to estimate the number of trips originating in each zone, i.e. *trip-end productions*, and the number of trips destined for each zone, i.e. *trip-end attractions*. The differences are well illustrated in the simple three-zone model in Fig. 6.4.

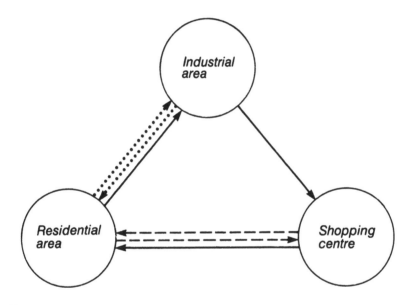

Key

••••••• Movement by first traveller, i.e. to work, then back home.

— — — Movement by second traveller, i.e. to the shops, then back home.

———— Movement by third traveller, i.e. to work, then to the shops, then back home.

Definition:

3 round trips

7 trips

14 trip ends (TEs), i.e. 6 residential TEs (3 productions, 3 attractions)
                                        4 employment TEs (2 productions, 2 attractions)
                                        4 shopping TEs (2 productions, 2 attractions)

Figure 6.4 *The definition of trips.*

Table 6.1   *Passenger car units*

Class of vehicle	Equivalent value in passenger car units (pcu)			
	Urban standards	Rural standards	Roundabout design	Traffic signal design
Private car, taxi, motor cycle combination, light goods vehicle (up to 30 cwt (1.5 tonnes) unladen)	1.00	1.00	1.00	1.00
Motor cycle (solo), motor scooter, moped	0.75	1.00	0.75	0.33
Medium or heavy goods vehicle (over 30 cwt (1.5 tonnes) unladen), horse-drawn vehicle	2.00	3.00	2.80	1.75
Bus, coach, trolley bus, tram	3.00	3.00	2.80	2.25
Pedal cycle	0.33	0.50	0.50	0.20

Three individuals live in the residential zone: one goes to work and then returns home; one goes shopping, and then returns home; and one goes to work, after work goes shopping and returns home. Depending on how a trip is defined, these movements constitute 3 round trips, 7 trips, and 14 trip ends. Readers should note the *directional symmetry* of travel, whereby the number of trips entering a zone tends to equal the number leaving in the course of a given day.

In earlier transport studies it was not uncommon to model trips on a combined basis, that is, to include all trips irrespective of purpose. However, it is now more usual to disaggregate the analysis and model travel patterns by trip purpose. For example, it is usual to distinguish:

1   trips between home and work place, i.e. *home-based work trips* (HBW);
2   trips between home and shops, educational establishments and other recreational and social activities — these can be treated separately, but are frequently linked together as *other home-based trips* (OHB);
3   *non-home-based trips* (NHB).

It is now necessary to address the question of what is making the trip by defining, more precisely, the traffic in the system. Appropriate measures of traffic are counts of the number of pedestrians, passengers, vehicles, or even freight tonnage and, depending on the design of the overall study process, trip generation models can be derived for person or vehicular movements. Whichever measure is used, ultimately it is necessary to translate the figure into vehicles to determine the fit between the vehicular capacity of a route and the vehicles demanding space of it. Vehicles of different types, of course, require different amounts of road space because of variations in size and performance and, in order to allow for this in capacity measurements, traffic volumes are often expressed in *passenger car units* (pcu). For example, to facilitate aggregation, the Department of the Environment (1973) has produced a classification which, assuming that the basic unit is the car, adjusts the weighting for other classes of vehicle according to the varying degrees to which they affect roads and junctions. The appropriate values for different types of vehicle under varying conditions are reproduced in Table 6.1.

Finally, a time period must be assumed. In highway practice the most widely used measure is *average annual daily traffic* (AADT), which counts trips in both directions and may, therefore, lead to confusion with the double counting of trip ends. An alternative to daily units is peak-hour volume, which can be measured directly or, as is often the case, is simply estimated as a percentage of AADT.

Having sorted out the basic definitional problems, a range of trip generation techniques can now be reviewed.

(a)   Land-use ratio methods
Land use is a particularly convenient way of classifying trip generating activities because it is easily measured and, given a well-established land-use planning framework such as in Britain, can be predicted with a fair degree of accuracy. As would be expected, different land uses produce different trip generation characteristics: for example, a commercial use normally generates more trips than open space. Furthermore, a given use itself generates differing numbers of trips depending on the intensity of development: for example, a given parcel of residential land produces very different characteristics, depending on whether it is developed at a high or low density. Although the range of land uses, especially in urban areas, can be very extensive, for trip generation studies only the most significant are considered. The method first establishes a set of land-use categories for the most significant uses, then, using *origin and destination* (O and D) survey data, trip origins associated with each category are counted and an average rate, for example, trips per acre, calculated for each land use. These rates are then used together with land-use forecasts to estimate future trip generation for each zone, it being assumed that there is a statistically significant variation in

mean rates among land-use categories and that these differences are stable over time from zone to zone.

The most significant land use is residential, since between 80% and 90% of all journeys have an origin or destination in the home. The actual measure of residential development used varies with the type of study being undertaken. For example, it can simply be measured in terms of acres of residential land in which case residential trip origins are stated as so many per acre. However, to reflect more accurately the intensity of development, more common measures are numbers of dwelling units, numbers of dwellings per acre, numbers of persons per acre, or even total population.

Commercial and industrial land uses are the next most important categories and it is usual to distinguish between retail and wholesale distribution, manufacturing and service industry, and office employment. Once again, the land area given over to each use could be measured and trip generation rates determined accordingly. But, as before, intensity of development is reflected more accurately using other measures such as amount of floorspace or numbers employed per unit of land area, for example, for shopping areas trip origins are often stated as so many per square foot of retail floorspace; similarly, manufacturing trips are usually stated as so many per employee or 100 employees.

Other significant land-use categories for which trip generation rates might be needed are educational and recreational developments. Numbers in attendance is the most common measure of the intensity of development at educational establishments, whereas recreational measures vary according to the facility or facilities in question.

(b)   Multiple regression models
Multiple regression analysis has been widely used in trip generation studies, particularly within the residentially based trip generation category. Regression techniques have a number of advantages over land-use ratio methods, not least the move towards a behavioural unit of analysis which focusses on households — after all it is households or parts of households that make trips, not acres or square feet. Home-based trips by purpose can be assembled as the dependent variables and data on the independent variables — including vehicle ownership, economic and social status of family units, household structure, density, distance from the CBD and so on — can be derived from the home interview and other surveys carried out in the preliminary stages of the study. As would be expected, most studies suggest that car ownership and household size are the crucial variables in explaining variations in the number of home-based trips.

The strengths and limitations of regression analysis are discussed in Appendix A.9, and it would be inappropriate to repeat these here, but readers should note the two important prerequisites before multiple linear regression can be applied: first, that a linear relationship exists between the dependent and independent variables; and, second, that the influence of the independent variables is additive — that is, the inclusion of each independent variable contributes towards accounting for the value of the dependent variable. Although multiple regression equations are widely used in trip generation studies for predictive purposes, a number of commentators have demonstrated how the assumed relationships are not, in fact, linear, and the assumptions of regression analysis are often not met, with the result that predictions are biassed.

(c) Category analysis

Category analysis, which was developed in Britain by Wooton and Pick (1967), attempts to overcome some of the problems associated with the use of multiple regression techniques and is based on the use of trip rates by household category rather than by zone, the results being aggregated later to give zonal trips. Wooton and Pick assumed that only three factors are of any real importance in affecting the amount of travel a household generates. Households are, therefore, cross classified according to these factors — i.e. household size and structure, household income, and car ownership — and mean trip generation rates for each category calculated from survey data. Assuming that trip generation rates for different categories of household will remain constant in the future, zonal traffic can then be estimated by simply multiplying the mean trip rates for each category by the number of households in that category for each zone and summing for all categories in the zone. For example, using census and other data it might be predicted that there were 100 households in a given zone, each owning one car, with one employed and one non-employed adult, and with gross income in the range £8000 to £10 000 per annum. Using data compiled as part of the initial household interview survey, it has already been estimated that households within this category make an average of 0.93 trips per 24 hour day, giving an estimated total zonal trip generation of 93 trips for these households.

Wooton and Pick's original analysis was based on six levels of income which, together with three levels of car ownership and six sets of household structure, defined a total of 108 household categories. With the use of data from two earlier UK transport studies, a trip rate for six different trip purposes and three modes was then calculated for each category. The estimated trip rates for each household classification from the two studies compared favourably. Subsequent studies have expanded the number of categories by adopting a wider range of household characteristics.

6.3.2 Trip distribution

Having estimated the number of trip ends generated by each zone, the function of trip distribution is to calculate the number of trips between one zone and another. For example, given the previously determined number T_i of trip ends generated in zone i, the trip distribution model calculates how many of these trips will travel to zone j and forecasts such interzonal transfers for every zone, i.e. it estimates the value of T_{ij} in the sequential traffic forecasting scheme illustrated in Fig. 6.1. In a study area containing n zones, there are n^2 possible zone-to-zone movements of this kind, and these are best described in a matrix of trips where each cell in the matrix represents the trips between two zones. The outcome of the trip distribution stage is, therefore, the production of such trip matrices, a simple example of which is illustrated in Table 6.2 (actual studies, of course, involve many more zones and hence considerably more trip combinations).

The procedures employed to make trip distribution forecasts all embody, either explicitly or implicitly, certain underlying principles. These include the familiar gravitational notion of movement being a function of attractiveness and an inverse function of friction due to distance. Although a number of different models have been developed, these tend to be more detailed examples of two basic methods: growth-factor methods and synthetic models.

(a)   Growth-factor methods
Models of this kind take the observed base-year interzonal flows and apply a growth factor to predict future design-year flows. Hence, in general terms, we can write:

$$T_{ij} = gt_{ij} \qquad (6.1)$$

where T_{ij} = future total trips between i and j
   t_{ij} = existing number of trips between i and j
   g = growth factor.

Early transport studies employed a uniform growth factor of this kind for the entire study area and then applied this to all interzonal transfers. Quite clearly, this can lead to substantial error when there are significant changes in the land-use pattern between the base- and design-years, affecting the distribution of population, commerce and industry. The family of growth factor methods can, therefore, be distinguished by the way they address this issue of deriving an appropriate growth rate and the four general types — uniform factor, average factor, Fratar and Detroit methods (reviewed in Finney, 1972) — are simply successive elaborations of the crude formula given by equation (6.1) above. For example, the Detroit method says that trips from zone i will increase by a rate F_i and will be attracted to zone j in proportion to the relative growth rate of trip making at j compared to the study area as a whole, or F_j/F, i.e.

$$T_{ij} = \left(t_{i \to j} \frac{F_i F_j}{F} + t_{j \to i} \frac{F_j F_i}{F} \right) \qquad (6.2)$$

where
$$T_{ij} = \text{future total trips between } i \text{ and } j \text{ (both directions)}$$
$$t_{i \rightarrow j} = \text{existing trips from } i \text{ to } j \text{ (one way)}$$
$$F_i = T_i/t_i$$
$$T_i = \text{projected trip-end productions at } i$$
$$t_i = \text{existing trip-end productions at } i$$
$$F_j = T_j/t_j$$
$$F = \frac{T}{t} \text{ (study area as a whole).}$$

In other words, the model applies the first of the underlying principles that movement is a function of attractiveness and that travel between two points (from i to j) will increase with the demand for travel from the first point (F_i measures the growth rate for trip making at i) and/or with an increase in the opportunities derived from travel to the second (the quotient F_j/F measures the change in the attractiveness of j). The converse is also true for movement in the opposite direction from j to i. Summing the right-hand side of equation (6.2) gives:

$$T_{ij} = t_{ij} \frac{F_i F_j}{F} \tag{6.3}$$

The following simple example demonstrates how equation (6.3) is used in practice.

Assume a three-zone study area in which the existing pattern of interzonal transfers is as illustrated in Fig. 6.5 and Table 6.2 (the flows are derived from the initial O and D survey). Projected trip ends have been derived as part of the trip generation exercise and these are also included in Table 6.2. Given these t and T values for each zone, F values can be derived, and equation (6.3) can now be applied as follows.

Table 6.2 *Existing trip distribution matrix*

From \ To	Existing flows			Existing study area totals derived from O and D survey t_i	Design-year totals estimated from trip generation analysis T_i	Growth factors $F_i = \dfrac{T_i}{t_i}$
	Z_1	Z_2	Z_3			
Z_1	100	250	150	500	1500	3
Z_2	250	200	100	550	2200	4
Z_3	150	100	150	400	800	2
				$t = 1450$	$T = 4500$	$F = 3.1$

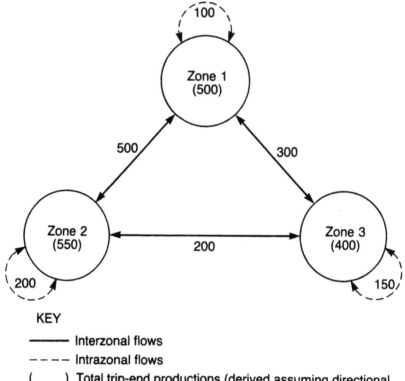

Figure 6.5   *Traffic flows in the three-zone study area.*

Given $T_{ij} = t_{ij} \dfrac{F_i F_j}{F}$

then

$$T_{11} = \frac{100(3)(3)}{3.1} = 290.3$$

$$T_{12} = \frac{500(3)(4)}{3.1} = 1935.5$$

$$T_{13} = \frac{300(3)(2)}{3.1} = 580.6$$

$$T_{22} = \frac{200(4)(4)}{3.1} = 1032.3$$

$$T_{23} = \frac{200(4)(2)}{3.1} = 516.1$$

$$T_{33} = \frac{150(2)(2)}{3.1} = 193.5$$

Total $= 4548.3$

Assuming directional symmetry of flows, a trip distribution matrix for the design-year can now be constructed as illustrated in Table 6.3.

Table 6.3 *Design-year trip distribution matrix (first iteration)*

To From	Design-year flows			Design-year totals T_i
	Z_1	Z_2	Z_3	
Z_1	290.3	967.8	290.3	1548.4
Z_2	967.8	1032.2	258.1	2258.1
Z_3	290.3	258.1	193.5	741.9
				$T' = 4548.4$

Readers will have noted that, as a consequence of the arithmetic, the number of trips resulting from the distribution exercise, 4548.4, does not match the number actually estimated as part of the trip generation process, 4500. Adjustments need to be made and, in the first instance, this involves the construction of a set of correcting growth factors which are derived as follows:

$$F_i = \frac{T_i}{T'_i} \quad \text{and} \quad F' = \frac{T}{T'} \tag{6.4}$$

The adjustment factors, which are listed in Table 6.4, are then applied to the principal formula (equation (6.3) as amended in (6.5)) to produce a revised second round of trip distribution estimates.

Table 6.4 *Adjusted growth factors*

	Design-year totals		Adjusted growth factors $F'_i = \frac{T_i}{T'_i}$
	Estimates from trip generation analysis T_i	Estimates from first round trip distribution analysis T'_i	
Zone 1	1500	1548.4	0.97
Zone 2	2200	2258.1	0.97
Zone 3	800	749.9	1.08
	$T = 4500$	$T' = 4548.4$	$F' = 0.99$

Hence, using the corrected growth factors, the revised trip distribution estimates will be as follows:

Given $T'_{ij} = T_{ij} \dfrac{F'_i F'_j}{F'}$ (6.5)

then $T'_{11} = \dfrac{290.3(0.97)(0.97)}{0.99} = 276$

$T'_{12} = \dfrac{1935.6(0.97)(0.97)}{0.99} = 1840$

$T'_{13} = \dfrac{580.6(0.97)(1.08)}{0.99} = 614.4$

$T'_{22} = \dfrac{1032.2(0.97)(0.97)}{0.99} = 981$

$T'_{23} = \dfrac{516.2(0.97)(1.08)}{0.99} = 546.2$

$T'_{33} = \dfrac{193.5(1.08)(1.08)}{0.99} = 228$

$T'' = 4485.6$

Once again, assuming directional symmetry of flows, a revised trip distribution matrix can be constructed and is illustrated in Table 6.5.

Table 6.5   *Design-year trip distribution matrix (second iteration)*

From \ To	Design-year flows			Design-year totals T''_i
	Z_1	Z_2	Z_3	
Z_1	276	920	307.2	1503.2
Z_2	920	981	273.1	2174.1
Z_3	307.2	273.1	228	808.3
				4485.6

As before, the number of trips predicted in the trip generation analysis does not appear to have been distributed. However, this second round of trip distribution estimates is certainly closer to the required figure than the first. In actual studies, interzonal growth factors are revised iteratively until a balance to within an acceptable limit, say plus or minus 5%, is achieved.

The simplicity and flexibility of growth-factor models made them very attractive in early transport studies. However, they are now rarely used because of a number of limitations. An important disadvantage is the necessity for comprehensive, and hence expensive, origin and destination surveys with high sampling sizes, in order to estimate smaller zone-to-zone movements. A second problem is their unreliability in predicting patterns of movement in areas where substantial land-use changes take place. However, the most important and fundamental criticism of growth-factor methods is their failure to adequately embrace the principle that movement is not only a function of attractiveness, but is also a function of friction due to distance. The failure to deal adequately with travel resistance implies that network changes have no impact on travel demand, and this is obviously not the case: for example, the construction of a motorway between two zones clearly leads to marked changes in travel behaviour. Synthetic models overcome some of these deficiencies by relating travel behaviour to both the generation and attraction of trips, and the resistance to travel within the network.

(b) Synthetic models
Synthetic models attempt to understand the actual causal factors underlying movement patterns in order to establish quantifiable relationships between trips and measures of attraction, generation and travel resistance. In this way, synthesis not only facilitates trip prediction, but also allows analysts to synthesize base-year traffic flows, thus avoiding the expense of having to comprehensively survey every cell in the trip matrix. The family of synthetic models include opportunity models, regression models and electrostatic models, and these are well reviewed elsewhere (see, for example Finney, 1972). However, perhaps the most widely used and extensively developed synthetic method of trip distribution employs a gravity model which explicitly embodies the familiar notion that trip making is a function of attractiveness and an inverse function of friction due to distance. The gravity concept was discussed in some detail in Chapter 5, and in trip distribution studies the model typically takes the form:

$$T_{ij} = T_i \frac{T_j K_{ij} F_{ij}}{\sum_{j=1}^{n} T_j K_{ij} F_{ij}} \tag{6.6}$$

where T_{ij} = the number of trips from i to j (one directional)
  T_i = projected trip-end productions at i
  T_j = projected trip-end attractions at j
  F_{ij} = an exponentially determined impedance factor expressing the average area-wide effect of spatial separation that approximates $1/D_{ij}^{\lambda}$ where D_{ij} is the actual measure of spatial separation
  K_{ij} = a zone-to-zone adjustment factor to account for other socio-economic influences on travel behaviour not already accounted for in the model.

Readers should note that equation (6.6) is precisely the form taken by the constrained gravity model, i.e. equation (5.17), derived earlier in Chapter 5.

As was suggested in Chapter 5, the measure of spatial separation D_{ij} used in the determination of the impedance factor can take a number of forms, including either straight line distance or travel time. However, in recent studies, the most commonly used measure is the perceived interzonal *generalized cost*, that is what a traveller thinks it costs him or her to travel between a pair of zones. For a motorist the cost function might include driving time costs, vehicle operating costs, parking charges, and access costs (cost of time spent walking from car/car park to actual destination and, where applicable, from home to car). For a public transport user the cost function would include access costs (cost of time spent travelling from home to bus stop and from bus stop to actual destination), waiting time costs, travel time costs, bus fares, and interchange and waiting time costs where more than one vehicle is involved (in such cases further allowances will need to be made for the additional travel time, fares, etc.). In general terms, therefore, a typical generalized cost function for a specific mode can be expressed thus:

$$C_{ij} = b_1 t_{ij} + b_2 e_{ij} + b_3 d_{ij} + P_j (+\alpha) \qquad (6.7)$$

where C_{ij} = generalized cost
   t_{ij} = travel time from i to j
   e_{ij} = excess time (access, waiting, etc.)
   d_{ij} = distance from i to j
   P_j = terminal cost at destination j
   α = a calibrating statistic which, in a sense, measures the inherent modal handicap
b_1, b_2, b_3 = constants, the value of which are derived through calibration.

Generalized cost is expressed in monetary units. It can, however, also be expressed in units of equivalent time — called generalized time — and in recent practice it is generalized time which has been used in the modelling of trip distribution and modal split to make traffic forecasts, with generalized cost tending to be used in the evaluation of alternative transport plans. Whichever measure is used, zone-to-zone travel resistances can be derived by building journey trees linking zone centroids, using Moore's algorithm as outlined in Section 6.3.4. Given an appropriate value for λ (λ is estimated using the calibration techniques described in Chapter 5), a zone-to-zone deterrence matrix can then be produced for use in the distribution process.

In smaller studies involving areas with populations of less than 100 000, the zone-to-zone adjustment factor is not necessary. However, for larger areas it has to be included to eliminate any 'systematic' errors and can be calculated by first preparing a trip distribution table of current behaviour based on observed true values t_{ij} and then comparing these with a modelled distribution of current interzonal transfers T_{ij}, according to the following formula:

$$K_{ij} = R_{ij} \frac{1 - X_i}{1 - X_i R_{ij}} \tag{6.8}$$

where R_{ij} = ratio of observed current interzonal transfers t_{ij}, to modelled current interzonal transfers T_{ij}

X_i = ratio of observed trips t_{ij}, to total observed trips leaving zone i.

The use of the gravity model can now be demonstrated, once again, by using a simple three-zonal model. Table 6.6 lists the projected trip-end productions and attractions for each zone together with the interzonal measures of spatial separation (generalized time or whatever) as listed in the bottom right-hand corner of each matrix cell. The problem is to find the upper left values in the matrix. Assuming $\lambda = 2$ and that the area is small and it is possible, therefore, to ignore K_{ij}, the basic equation (6.6) becomes:

$$T_{ij} = \frac{T_i\, T_j\, F_{ij}}{\sum\limits_{j=1}^{n} T_j\, F_{ij}}$$

$$= \frac{T_i\, (T_j/D_{ij}^2)}{\sum\limits_{j=1}^{n} (T_j/D_{ij}^2)} \tag{6.9}$$

Equation (6.9) can now be used together with the data in Table 6.6 to derive the following pattern of interzonal transfers.

Given $T_{ij} = \dfrac{T_i\, (T_j/D_{ij}^2)}{\sum\limits_{j=1}^{n} (T_j/D_{ij}^2)}$

then $T_{11} = \dfrac{750(750/2^2)}{750/2^2 + 1100/5^2 + 400/3^2} = 509.6$

$T_{12} = \dfrac{750(1100/5^2)}{750/2^2 + 1100/5^2 + 400/3^2} = 119.6$

$T_{13} = \dfrac{750(400/3^2)}{750/2^2 + 1100/5^2 + 400/3^2} = 120.8$

$T_{21} = \dfrac{1100(750/6^2)}{750/6^2 + 1100/4^2 + 400/3^2} = 171.0$

cont.

Table 6.6   *Design year trip-end constraints and deterrence matrix*

From \ To	Z_1	Z_2	Z_3	Design-year totals T_i
Z_1	T_{11} $D_{11} = 2$	T_{12} $D_{12} = 5$	T_{13} $D_{13} = 3$	750
Z_2	T_{21} $D_{21} = 6$	T_{22} $D_{22} = 4$	T_{23} $D_{23} = 3$	1100
Z_3	T_{31} $D_{31} = 2$	T_{32} $D_{32} = 2$	T_{33} $D_{33} = 1$	400
Design-year totals T_j	750	1100	400	$T = 2250$

$$T_{22} = \frac{1100(1100/4^2)}{750/6^2 + 1100/4^2 + 400/3^2} = 564.2$$

$$T_{23} = \frac{1100(400/3^2)}{750/6^2 + 1100/4^2 + 400/3^2} = 364.8$$

$$T_{31} = \frac{400(750/2^2)}{750/2^2 + 1100/2^2 + 400/1^2} = 87.0$$

$$T_{32} = \frac{400(1100/2^2)}{750/2^2 + 1100/2^2 + 400/1^2} = 127.5$$

$$T_{33} = \frac{400(400/1^2)}{750/2^2 + 1100/2^2 + 400/1^2} = 185.5$$

The results of these calculations are recorded in Table 6.7.

The property of the singly constrained (in this case production-constrained) gravity model is that while a trip-end balance exists at the origin end so that row totals match those specified for the T_is, the job-end summations need not match the respective T_js. As with growth-factor methods, the initial results may not, therefore, be satisfactory. For example, in Table 6.7 the distributed arrivals $T_j(d)$ are clearly not equal to the generated arrivals T_j used in the first instance in Table 6.6. Once again, the situation may be improved by iteratively applying an adjustment factor, T_j', until a satisfactory balance is achieved. T_j' is computed as follows:

$$T_j' = \frac{T_j}{T_j(d)} \, , T_j \qquad (6.10)$$

Table 6.7 *Design-year trip distribution matrix (first iteration)*

To / From	Design-year flows			Design-year totals $T_i(d)$
	Z_1	Z_2	Z_3	
Z_1	509.6	119.6	120.8	750
Z_2	171	564.2	364.8	1100
Z_3	87	127.5	185.5	400
Design-year totals $T_j(d)$	767.6	811.3	671.1	

The need to apply such adjustment factors can be avoided by using a doubly constrained gravity model where one not only fixes the number of trips generated from the origin zones, but also limits the number attracted to each destination zone (see, for example, Blunden and Black, 1984, pp. 54–59, or Foot, 1981, pp. 87–98).

The practical and operational limitations in applying the gravity concept were discussed in Chapter 5. Notwithstanding these, gravity models have been widely used in trip distribution studies.

6.3.3 Modal split

The purpose of modal-split analysis is to determine what proportion of trips will be made by public transport as opposed to private car. A number of different approaches have been developed to estimate the split and all are based, not unreasonably, on the assumption that modal choice depends on a traveller's perception of the merits of each mode of transport in relation to its competitors. As was suggested earlier, the position of modal-split analysis in the sequential traffic forecasting scheme will vary from study to study and, indeed, it is not uncommon to handle it in conjunction with one of the other stages such as separately generating trips by mode, etc.

(a) Modal split with trip generation

In the earliest transport studies the analysis of modal split was often introduced at the generation stage with modal choice prediction based on a number of factors. A typical model might proceed as follows. First, forecast future person trips for each zone as part of the normal trip-generation exercise; second, for each zone separate CBD trips from non-CBD trips; third, using historic data, estimate the proportion of CBD trips that will be made by public transport for each zone — the balance will be made by car; and finally, for non-CBD trips build regression equations to estimate public transport trips for other trip purposes using factors such as residential density and cars per household as independent variables. For each zone, therefore, it is possible to derive:

1   public transport trips to the CBD;
2   public transport trips to school;
3   'other' public transport trips;
4   the residual of private car trips.

(b)   Modal split with trip distribution

When modal split is introduced at the trip distribution stage, the purpose is to determine the proportion of interzonal transfers that will travel by each mode. The procedure involves allocating journey movements to different modes after total person trips between pairs of zones have been distributed. A typical study would take the initial aggregate zone-to-zone pattern of movement derived as part of the standard trip distribution exercise and then, using regression techniques similar to those mentioned above, calculate the proportion of trips that are likely to use public transport. The remainder obviously travel by car. Instead of a single trip distribution matrix, therefore, the product of this form of analysis is two matrices, one for each mode.

(c)   Diversion curves

Another method of modelling the modal split is to use diversion curves. These are empirically determined relationships constructed from survey data on the usage of alternative modes. For example, the London Transportation Study used very simple diversion curves incorporating only one variable, that is, the percentage of public transport trips between any pair of zones was related to the travel-time ratio for three trip purposes. In Fig. 6.6, therefore, given a travel-time ratio of 1.45 by the competing modes for a particular zonal pair, we can see that 40% of non-home-based trips between the zones will be by public transport and, by implication, 60% by car.

6.3.4   Assignment

Having distributed predicted person trips by mode, the next stage in the sequential traffic forecasting scheme is to simulate route choice. It is first necessary, however, to convert person trips to vehicle trips, and this is achieved by the simple application of average car and bus occupancy rates (usually differentiated by trip purpose). The traffic assignment process then involves the allocation of this given set of vehicle trip interchanges to the actual transport system. It requires as a basic input a description of the existing or proposed transport network. Individual routes through the network are then defined and interzonal trip movements loaded onto the paths selected.

In all assignment procedures it is assumed that trips have their origins and destinations at zone centroids and usually select routes so as to minimize the

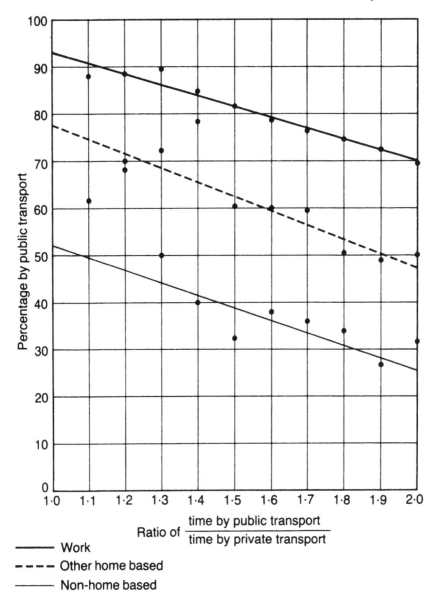

Figure 6.6 *London Transportation Survey modal-split diversion curve.*

generalized cost of travel for particular journeys. The least-cost route from one zone to another is known as a tree and assuming that the generalized cost of travel has been established for each link in the network, it is possible to build *least-cost journey trees* from one zone to another using Moore's (1957) algorithm.

(a)   Moore's algorithm

Moore, who was actually working in the field of telecommunications, developed the procedure to solve the problem of routing the direct dialling of long-distance telephone calls. But the method was adjusted for use in traffic assignment and the basic process can be illustrated with the aid of Fig. 6.7. The diagram assumes a three-zone study area and the given network identifies the three zone centroids Z_1, Z_2 and Z_3. The problem is to build a least-cost journey tree for centroid 1.

Starting at Z_1, go to each connecting node and record the generalized cost of travel to it, i.e. $Z_1 \rightarrow 1 = 10$ and $Z_1 \rightarrow 2 = 10$. The node closest (lowest cost) to the home node (the centroid) is considered next. In this case,

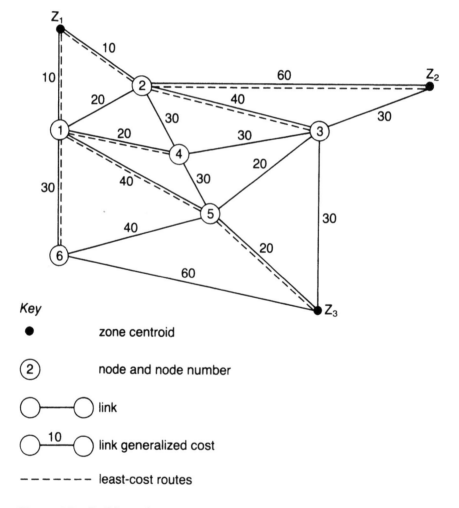

Key

●	zone centroid
②	node and node number
○——○	link
○—10—○	link generalized cost
− − − − − −	least-cost routes

Figure 6.7   *Building a least-cost journey tree for centroid 1.*

however, nodes 1 and 2 are equidistant from Z_1, so the lowest numbered node is used, i.e. node 1. The cumulative cost from Z_1 through node 1 to all nodes directly connected to node 1 is now recorded, i.e. $Z_1 \to 1 \to 2 = 30$, $Z_1 \to 1 \to 4 = 30$, $Z_1 \to 1 \to 5 = 50$ and $Z_1 \to 1 \to 6 = 40$. As can be seen, two routes have been identified to node 2 and the least cost $(Z_1 \to 2 = 10)$ is chosen while the other $(Z_1 \to 1 \to 2 = 30)$ is rejected. Node 2 is now considered, since it is the next closest to Z_1, and the cumulative cost from Z_1 through node 2 to all nodes directly connected to node 2 is recorded, i.e. $Z_1 \to 2 \to 3 = 50$, $Z_1 \to 2 \to 4 = 40$, and $Z_1 \to 2 \to Z_2 = 70$. Once again, two routes have now been identified to node 4 and the least cost $(Z_1 \to 1 \to 4 = 30)$ is chosen whilst the other $(Z_1 \to 2 \to 4 = 40)$ is rejected. This process is repeated until all nodes have been reached via the minimum path from centroid 1 (see Table 6.8), and this defines the least-cost journey tree. Clearly a whole series of such decision trees can be built and these are then used to assign the trip matrix to the various links in the network.

Table 6.8 *Building a least-cost journey tree for centroid 1*

Link	Cost	Link combinations	Total cost	Least-cost routes
$Z_1 \to 1$	10			least cost to node 1
$Z_1 \to 2$	10			least cost to node 2
$1 \to 2$	20	$Z_1 \to 1 \to 2$	30	
$1 \to 4$	20	$Z_1 \to 1 \to 4$	30	least cost to node 4
$1 \to 5$	40	$Z_1 \to 1 \to 5$	50	least cost to node 5
$1 \to 6$	30	$Z_1 \to 1 \to 6$	40	least cost to node 6
$2 \to 3$	40	$Z_1 \to 2 \to 3$	50	least cost to node 3
$2 \to 4$	30	$Z_1 \to 2 \to 4$	40	
$2 \to Z_2$	60	$Z_1 \to 2 \to Z_2$	70	least cost to zone 2
$4 \to 3$	30	$Z_1 \to 1 \to 4 \to 3$	60	
$4 \to 5$	20	$Z_1 \to 1 \to 4 \to 5$	50	
$5 \to 3$	20	$Z_1 \to 1 \to 5 \to 3$	70	
$5 \to Z_3$	20	$Z_1 \to 1 \to 5 \to Z_3$	70	least cost to zone 3
$6 \to 5$	40	$Z_1 \to 1 \to 6 \to 5$	80	
$6 \to Z_3$	60	$Z_1 \to 1 \to 6 \to Z_3$	100	
$3 \to Z_2$	30	$Z_1 \to 2 \to 3 \to Z_2$	80	
$3 \to Z_3$	30	$Z_1 \to 2 \to 3 \to Z_3$	80	
$Z_2 \to 3$	30	$Z_1 \to 2 \to Z_2 \to 3$	100	

(b) All-or-nothing assignment
When all the trips for a particular interzonal transfer are allocated to the links forming the least-cost path between two zone centroids, this is known as all-or-nothing assignment and, at this stage in the modelling process, it is not unreasonable to load all vehicle trips onto these least-cost routes.

However, experience has shown that all-or-nothing techniques, while appropriate for public transport assignments, can lead to poor results for other traffic, and this has led to the development of capacity restraint methods.

(c)   Capacity restraint assignment

Capacity restraint acknowledges that traffic joins the network gradually, and links are therefore loaded incrementally. As this loading takes place, the increasing number of vehicles slows the traffic flow on the link in question making it less attractive, and other nearby links begin to assume the role of least-cost alternatives.

In its most basic form, assignment with capacity constraint simply sets an absolute capacity on each link and then assigns in an all-or-nothing manner, filling to capacity the least-cost routes first and then moving to the next best and so on. More complex methods depend on an undestanding of the relationship between the volume of traffic using a link and the speed at which that traffic can move (the speed-flow relationship) and involve iteratively amending the least-cost journey trees for each increment of load, that is, as traffic volumes increase on attractive routes, so the free flow of traffic is limited, causing average speed to drop and consequently increasing the travel cost on that link. A typical impedance function is illustrated in Fig. 6.8 where, if the loading moved a link's volume/capacity ratio from point A to point B, this would have the effect of reducing average speeds from 30 to 20 miles per hour (50 to 30 km/h).

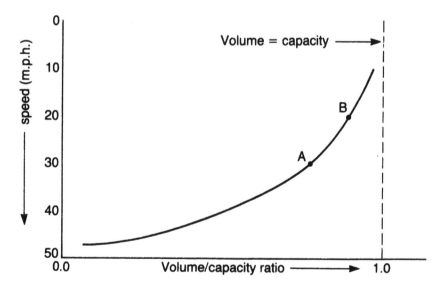

Figure 6.8   *The relationship between speed, volume, and capacity.*

6.4 Evaluation

The major reason for developing a transport model is not just to predict traffic flows, but also to assess the limitations of the transport network by assigning estimated future trips to the existing system (including already committed improvements). Deficiencies will be identified and a range of alternative strategies can be developed to address these. For each strategy in turn, suggested improvements and extensions can then be evaluated by reassigning estimated future trips to the revised network which includes the proposed improvements. This kind of evaluation is simply technical, however, since all strategies might prove operationally feasible and able to deal more than adequately with network deficiencies. It is still necessary to decide which strategy should be preferred, or whether to do nothing (the *do-nothing option* represents the travel conditions likely to prevail in the design-year if no additional major capital projects are committed — it is the strategy from inspection of which new options should ideally be developed). In any event, it is necessary to examine thoroughly the possible consequences of all options and this evaluation is usually considered in four stages.

First there is *numerical evaluation*, which checks the computational validity of forecasts. The aim is to confirm that the results obtained from the sequential traffic forecasting scheme correctly reflect the input assumptions. The concern is with computational accuracy and data quality.

Operational evaluation involves questions of technical feasibility such as those mentioned above, that is, it checks to see if the network resulting from a proposed strategy adequately satisfies the forecast travel patterns. The concern is with traffic flows, travel speeds and the capacity limitations of different types of transport system.

The third stage of the evaluation process considers the environmental and aesthetic impact of each strategy. *Environmental evaluation* is becoming increasingly important as society begins to question the price we are prepared to pay for an efficient transport system. What is technically feasible and financially beneficial might well be quite unacceptable environmentally. But the question of actually quantifying these environmental disbenefits, particularly when they are essentially aesthetic and intangible, remains. Environmental evaluation is still in large measure subjective, a matter of judging what is desirable and acceptable.

Economic evaluation is the final stage in the process and attempts to measure, given the budget constraint, which strategy yields the greatest financial benefits. Although conventional investment appraisal methods have sometimes been used, it is now more usual to employ cost–benefit analysis. Cost–benefit techniques take a wider view in evaluating alternative strategies and probably negate the need to disaggregate the evaluation process into four stages. Environmental and, to a lesser extent, operational matters are usually included in the cost–benefit study. The strengths and weaknesses of cost–benefit analysis are discussed more fully elsewhere (see, for example, Department of Transport, 1972).

6.5 Limitations of transport studies

The fragility of many of the assumptions underlying the techniques for forecasting and evaluation in transport planning are largely self-evident. It is, however, worthwhile highlighting some of the more important limitations of the approach.

Notwithstanding that predictions in the level of economic activity are notoriously precarious, even at the national and regional level (see Chapter 4), the process begins by making forecasts of jobs and incomes for the study area at a level of aggregation which demonstrates excessive optimism, given the smallness of the unit of observation.

The forecast of travel demand, in the trip generation stage, then proceeds on the basis of a given/predicted pattern of land use, it being assumed that this pattern is constant between the strategies that will subsequently be developed and examined in the evaluation process. There is no acknowledgement that land use and locational decisions are likely to change because of the restructuring of accessibility resulting from network changes occasioned by the adoption of different strategies. Once again, it is forecasts of the volume and location of employment that are likely to be particularly sensitive to this issue.

The trip distribution stage either suffers from over simplicity when growth-factor methods or their equivalent are used, or, when gravity models are employed, from many of the limitations of the gravity concept that were discussed in Chapter 5.

Modal split is viewed as a matter of probabilistic choice between competing modes, but is this appropriate? Any forecast must be based on explicit and implicit assumptions about the policies of actual decision makers and modal choice, or more specifically the promotion of public transport represents an important variable that can be 'controlled' to contain the growth process. In other words, should the modal split be predicted or should it be taken as given, that is, defined as a policy constraint in the forecasting procedures?

The sequential traffic forecasting scheme considers assignment and ultimate route choice independently of destination choice. As O'Sullivan and others have noted, the implication is 'that the build-up of congestion in certain parts of the network will not influence the desirability of different residential areas or the attractiveness of particular work or shopping places.' (O'Sullivan *et al.*, 1979, p. 166). Experience and observation suggests otherwise.

It should also be noted that all the procedures discussed thus far have focussed on person journeys with no development of analogous procedures to analyse and forecast commercial vehicle journeys. Evans highlights this point when he notes that 'forecasting commercial vehicle movements often involves no more than the grossing up of an observed matrix of inter-zonal commercial vehicle journeys by a factor which might be based on a forecast growth of GNP' (Jones, 1977, p. 131). Given increasing public concern with the environmental consequences of heavy goods vehicles, this perfunctory treatment of an increasingly important class of vehicle movement is clearly unsatisfactory.

Although the modelling process ostensibly examines the interaction be-tween demand and supply, the sequential traffic forecasting scheme, as defined, clearly has only a single direction of causation from the demand for travel to the provision of facilities to meet that demand. And yet in reality, as some of the points raised above suggest, the overall system exhibits complex two-way relationships not catered for in such a sequence. Clearly, for example, the supply of transport infrastructure can have a profound effect on demand by generating additional traffic. To obtain reasonable estimates from the generation, distribution and modal split stages it is necessary to know T_{ijkl} for all i, j, and k — in other words some form of feedback mechanism is required to make allowances for the changes brought about in the supply conditions following assignment. For this reason, the initial forecast is best thought of as a first approximation which requires additional iterations to improve its accuracy and sensitivity to subsequent policy changes. In practice, however, such cyclical procedures are rarely evident.

These and other criticisms of traffic forecasting have led to the develop-ment of more behaviourally sound and policy-responsive models, known variously as behavioural models, individual choice models, or discrete-choice models. To a large extent, the traditional approach and the associated planning process were preoccupied with the planning and design of high-capacity road systems in urban areas. As can be seen from the preceding paragraphs, the future is perceived as a comfortably distant design year and system users are mere abstractions subsumed in measures of vehicular flow. However, as disenchantment with road building and increasing concern with transport system management has grown, so too has the context in which demand is forecast. An increasing concern with broader environmental issues, the energy crisis, and a whole range of other social, economic and political factors has shifted the emphasis of transport policy

> from the long run to the short, from large scale to small, from single mode to mix of modes, from facilities to services, and from vehicles to people . . . with this changing perspective in transport planning has come a reaffirmation of the importance of individual behaviour and individual values. (Stephen and Meyburg, 1976, p. XV)

The result is a series of models which focus on individual behaviour involv-ing choice among discrete alternatives — in the present context, the determinants of travel choice (see, for example, Domencich and McFadden, 1975 and Hensher and Johnson, 1981). These models, which are founded on the economics of consumer behaviour and the psychology of choice behaviour, are now an accepted planning tool in dealing with modal choice, although they have yet to make their pressure felt as a more general tool in transport analysis. They have an intellectual appeal which resulted in an extremely fruitful area of research in the 1970s, spawning a number of attractive theoretical constructs, but in practice the creation of workable models has proved more elusive. However, they do represent an approach to travel forecasting that appears to hold the greatest potential for improving the transport planning process. In the meantime, the sequential traffic forecasting scheme is likely to continue as the standard method, albeit in a modified form.

6.6 Conclusions

Transport planning is far from being an exact science, and yet the basic procedures developed within the overall transport model have survived remarkably well, given their obvious limitations. However, the procedures are not cheap, and in many cases potential accuracy has to be forgone to avoid excessive expense on initial surveys and subsequent data analysis and computation. Furthermore, the general approach tends to be more appropriate for large urban areas or counties — for smaller towns specific problems can be addressed in a more selective fashion. As with many of the techniques outlined in this text, the transport model is simply an aid to decision making. Its ultimate form must, as a result, be related to the types of decision that are to be made, but therein lies the ultimate weakness of traffic forecasting, namely its circularity. The sequential traffic forecasting scheme contains no constraints on the growth processes that are being forecast — the concern is with what is likely to happen, not what is desirable. But central and local governments can and should prescribe transport policy which must, of necessity, severely constrain traffic forecasts. Planning procedures need to move away from the rather mechanistic extrapolations which are traditionally characteristic of the highway engineer towards a more policy-responsive behavioural methodology.

7   Recreation

7.1 Introduction

Recreation is an area of activity which has been relatively neglected by land use planners, and of all the topics covered by this book, it is the one which is furthest from having a standard forecasting methodology. Recreation is not an easily defined homogeneous entity; rather it consists of a diversity of activities undertaken at a wide range of locations. It has been suggested (Burton, 1971) that it is easier to define what it is not: it is not work nor 'personal and social' obligations such as sleeping and washing. However, it is not an alternative to work, but a complement, as it presupposes work. The characteristic purposes of recreation are rest, amusement, personal and social development, disinterested education and improvement of skills. Particularly in the latter cases the distinction is problematic.

Some writers have tried to distinguish between recreation and leisure with the latter being more passive. However, Burton (1971, p. 20) maintains that 'Recreation, in its widest sense, is identical with leisure.' A more useful distinction is often between recreation and tourism, with the latter involving staying away from home and thus having implications for the provision of accommodation. Even here the division is not clear-cut and such distinctions depend on the nature of the topic of study. In National Parks the distinction may be of interest, but it is not so for provision of squash courts.

The consumption of recreation has been steadily growing with rising incomes, improvements in transport, increased car ownership, increased education, and increased time free from work. The structure of consumption has become more complex with each of these changes. More recently, rises in unemployment have further altered consumption patterns and requirements.

Burton (1971) suggests that recreation can be categorized by type, timing and location. An elaboration of this is shown in Table 7.1. It includes hiking in the countryside, playing squash at a local sports centre, visiting a theatre,

Table 7.1   *Recreation categorization*

Type	active/passive
	organized/unorganized
	solitary/group orientated
	requiring facilities/not requiring facilities
	involving skills/not involving skills
	involving costs/not involving costs
Timing	day visits/overnight stays
Location	indoors/outdoors
	urban/rural
	close to home/far from home
	close to work/far from work

library or museum. Most recreation time, however, is spent at home, watching television and gardening being the most popular. As non-work time is limited, all recreational activities are in competition with each other in some sense.

A large number of bodies are involved in the provision of recreational facilities. These can be divided into four broad categories: government (national and local), commercial enterprises, clubs and voluntary groups. Parks and libraries are typically publicly provided, while golf courses and tennis courts may be public or run as private clubs which charge a membership fee. By contrast, cinemas are primarily commercial enterprises while theatres may be public or private, but often require subsidies. The voluntary sector is involved in activities such as youth clubs. Access may thus be open to all, at a price or free, or open only to members.

Even within the public sector there is a wide range of bodies with responsibilities for different activities. A number of central government departments have responsibilities, as do numerous quangos ranging from the Sports Council to the Arts Council and the Forestry Commission. At the national level there is no integrating organizational structure, while at the local level recreation is often the function, not of the planning department, but of a separate recreation department.

Town planners have always had some concern with recreation through, for example, the provision of public open space. Green belts were also originally supposed to have a substantial recreational function. More recently the National Parks and Access to the Countryside Act 1949, the Countryside (Scotland) Act 1967 and the Countryside Act 1968 have all involved planners in recreation.

At the national level attempts have been made to plan for recreation, usually on an activity basis. At the local level, the wide variety of such activities has meant that planning policy has substantially different relevance to each. Planners are more involved in site-specific recreation and often in more rural locations. In general, recreation tends to be viewed as a social service, the provision for which is based on norms. One view even sees provision as helping to control 'anti-social behaviour'. Traditionally, local government provision has been heavily based on parks, playing fields and swimming pools.

Torkildsen (1983) suggests that, whereas planners have tended to ignore recreation and have concentrated on transport, housing and shopping, the leisure professions have ignored planning, being more interested in the management of facilities.

> The problem with planning for leisure is that, generally speaking, the planning profession knows very little about leisure while the leisure professions know very little about planning.
> (Quoted in Torkildsen, 1983; after Veal, 1982)

Structure plans, which are concerned with matters of strategic importance, rarely consider recreation provision in depth. Their main concern tends to be larger conservation and tourist sites in rural areas. Local plans also pay little attention to recreation. A 1979 survey of local plans in England and Wales (Henry, 1980) showed that 33% had no review of recreation facilities, and 20% had only a limited consideration. The survey concluded that even those plans which did consider the issue were characterized by a weakness in methodology, a lack of comprehensive coverage, a superficiality, and the application of outdated and inappropriate standards. Only where there was a separate recreation department with a research section was the analysis adequate. In Scotland, planning for recreation is rather more coordinated, with guidance from the Scottish Office through the National Planning Guidelines and a number of detailed national studies which provide a basis for recreation and tourism planning in structure and local plans. Nonetheless, this could not be claimed to be comprehensive.

7.2 Recreation demand

7.2.1 The structure of demand

In recreation literature demand has two meanings (Wilkinson, 1973). First, there is a popular usage referring to the number of visitors or users. More correctly this is consumption and it is this that many forecasting techniques measure. Second, there is the strict economic definition which relates the volume or quantity to price. Consumption is the interaction of demand with supply. If there is no price it is not possible to identify the demand–price relationship (Burton, 1971). In such circumstances changes in consumption represent supply changes or changes in consumers. Thus supply typically creates consumption (a similar effect was observed in Chapter 5 with the provision of new roads).

As many recreational activities have no admission price, and other costs to participants may not be regarded as costs (for example, travel costs), the concept of economic demand may have limited meaning. Demand is often used as a term to describe consumption with little or no reference to price. Such a usage will be adopted in this chapter unless it is explicitly stated otherwise.

It has been suggested (Burton, 1971; Gratton and Taylor, 1985; Roberts, 1974) that when considering recreation it is necessary to consider a number of components of demand. For this purpose, the population of an area may be divided into the following categories:

1   *Existing demand* — those currently using the facility. (This may be difficult to measure where there is no price or the activity is unconfined, such as walking.)
2   *Latent demand* — this can be referred to loosely as unrealized demand, and falls into two categories:
    (a)   *deferred* — those who want to participate and have the means and time but no facilities or else no knowledge of the facilities;
    (b)   *potential* — those who are without the means or time, but if conditions were right would take part.
3   *No demand* — those who do not participate through age, ill health or lack of interest.

Thus new facilities may have one or more of the following effects:

1   allow *latent* demand to be realized;
2   *induce* demand from those with no demand, by for example, creating interest;
3   *divert* demand from old to new facilities;
4   *substitute* demand from different types of activity or facility.

The provision of new facilities may, therefore, both change the *absolute volume* of demand and the *distribution* of demand. This may involve changes in the number of persons who participate and changes in the amount of participation of existing participants. When forecasting demand changes it is important to be aware of the above structure and the most important impacts on it.

7.2.2 Influences on demand

The demand for the use of facilities or for participation in activities is influenced by a complex number of factors. These may be conveniently divided into three categories (Cheung, 1972; Roberts, 1974).

1 *Factors relating to potential users as individuals*
   (a) the number of people in an area
   (b) their geographic distribution
   (c) their average income and the distribution of income
   (d) their average leisure time and its distribution
   (e) the age and sex distribution of the population
   (f) the socio-economic characteristics, such as occupation,
        education, knowledge of facilities, experience of activities
   (g) individual skill, motivation and preference.
   It should be clear that many of the above characteristics are strongly
   correlated, such as income and social class.

2 *Factors relating to the recreation facility or area*
   (a) the attractiveness of a facility or an area in terms of the quantity
        and quality of facilities
   (b) the management of the facility area
   (c) the capacity of the area or facility
   (d) the availability of substitutes or alternatives.
   As before these characteristics are likely to be correlated.

3 *Factors relating to the interaction of (1) and (2) above*
   (a) the distance or time or cost of travel between the residence or
        place of work and the area or facility. There is a pronounced
        'distance decay' effect which varies substantially for different
        activities. This is similar to shopping, but whereas, in some sense,
        shoppers must shop, people need not participate in a particular
        leisure activity. Thus, as distance increases not only does the
        proportion of the population using a facility decrease, but the
        frequency of use of individual users also decreases
   (b) the advertising of the area or facility.

These factors may affect the *generation* and the *distribution* of demand,
and thus the location as well as the level of use of the facilities. Planning for
recreation, therefore, involves not only assessment of future consumption
levels, but consideration of the spatial distribution of consumption and the
implications for provision. However, for some activities such as hill walking,
very little control can be exerted on location, and control is also more
difficult when a facility is provided by the private sector.

There are clearly observable interrelationships between demand and the
various factors listed above. It is, however, difficult to identify the structure
of causality. Moreover, the parameters are often highly unstable over time.
This is a particularly important problem in recreation forecasting, much
more so than in other forecasts. Recreation activities are notoriously prone
to changes in 'fashion' — the rise and fall of ten-pin bowling and the current
rise in popularity of squash are but two examples. Such volatility makes

forecasting, for anything other than a short period, particularly difficult for many types of recreation.

7.2.3 The development of forecasting

With the expansion of recreation during the 1950s and 1960s the need to plan increased, and with it the need to forecast. Burton (1971) suggests that there were four stages in the development of recreation forecasting. Initially there was a period of despair with the realization that many important causal variables were non-measurable. Even those which could be measured could not be forecast with any degree of accuracy and recreation forecasts were of no practical value. There then followed a period where estimates were little more than informed judgements based on information and experience. This was thought to be the best approach possible, but was a mixture of expectation and normative planning. The third stage was more optimistic and was based on extrapolation of trends. From this a fourth stage developed during which predictive models were constructed, adopted and adapted from other fields such as traffic engineering and econometrics. Since Burton offered this analysis the subject has developed, initially with the development of ever more elaborate models, the value of which, either to understanding or practice, is debatable. Recently there has been a more cautious approach to the construction and application of complex mathematical models.

Forecasts can cover one activity in one locality or a wide range of activities. They can be site-specific or for an entire 'market', or they can be long term or short term. Forecasts may be used to assess future demand for an activity or facility and so guide policies for provision, or to assess the impact of a particular proposal. In the private sector, demand forecasts may be used to assess the financial feasibility of a project. The wide variety of possible subjects makes comprehensive coverage impossible and generalization difficult. Moreover, the mathematics involved in many models is far beyond the scope of this book. The emphasis of this chapter is on the simpler variants which are likely to be of use to local authority planners. It should, however, be noted that many of the more complex models are elaborations of the basic structures presented here. Discussion is divided into three sections: normative planning; forecasting for activities; forecasting for sites or areas.

7.3 Normative planning

7.3.1 Standards

It is necessary to draw a distinction between a norm or a standard, which sets out what it is intended should happen, and a forecast of what is expected to happen under certain assumed circumstances. Traditionally, the planning of recreation has placed a strong emphasis on standards of provision. When

Table 7.2 *Recreation planning standards*

Facility	Standard
Playing fields	6 acres/1000 population
District indoor sports centres	One per 40 000–90 000 plus one additional per 50 000 (17 m² per 1000 population)
Indoor swimming pools	5 m²/1 000 population
Golf courses	One 9-hole course per 18 000

Sources: Veal (1982); Gratton and Taylor (1985).

standards are used, demand forecasting is, at best, an adjunct to recreation planning, rather than a fundamental prerequisite.

The heavy reliance on standards, such as those shown in Table 7.2, is the result of an earlier tradition of paternalism in provision, and the inadequacy of forecasting techniques. Forecasts are unlikely to be of value for periods of more than ten years and even then they may be unreliable. In many cases it is likely that the supply of a facility is more likely to affect consumption than changes in demand determinants. This can be explained by factors such as latent demand as outlined in Section 7.2 above. The relationship between consumption and supply is further complicated by the way in which current policies (such as provision of swimming lessons) influence future demand (for swimming pools).

Given these difficulties it is not surprising that planners have resorted to the use of nationally set standards. There are, however, problems as most standards are unresearched. Nonetheless, through constant use they have become widely accepted and little challenged. Typically they are applied to the whole population although consumption patterns and preference differ between social classes and from area to area.

The main advantage of normative-based planning of recreation is that it is likely to lead to action, and as most standards are conservative this is unlikely to lead to overprovision of facilities. Standards can be used to identify areas of relative underprovision, but even then a geographical spread of facilities or one based on facility hierarchies need not lead to equality of *access*. Nonetheless, with due consideration of local circumstances and flexibility, standards can provide a starting point for local authority policy in recreation provision. They are closely linked to *participation rates*, which are discussed in Section 7.4.1. One danger is that policies are influenced by expressed need and thus the more vocal groups receive attention.

7.3.2 Need
Planning based on standards uses a definition of need derived from policy

makers. Need may also be defined and measured in three other ways: felt, expressed and comparative. Felt need is based on surveys of people's attitudes and aspirations. It is generated in complex ways and may be prone to changes in fashion and advertising. Expressed need depends, for example, on pressure groups, and so is likely to overrepresent the better organized and more vocal. Comparative need considers patterns of provisions and distributions of facilities to identify *relative* deprivation. Each may be used as an input to policy. Note that although the terminology is different there are conceptual similarities with the terms used in housing (see Chapter 3).

The following two sections consider separately forecasting for particular activities and forecasting for individual sites. The distinction is, however, not clear-cut.

7.4 Forecasting for particular activities

This section examines techniques used for forecasting for particular activities or for whole markets. Methods are included which can be used at national, regional or local levels. Those considered are:

1   participation rates;
2   trend lines;
3   regression analysis;
4   the Delphi method.

7.4.1 Participation rates

The participation rate, as the name suggests, represents the proportion of persons who participate in an activity. Thus:

$$r = \frac{N}{P} \qquad\qquad (7.1)$$

so

$$N = rP \qquad\qquad (7.2)$$

where N = the number participating
P = the population
r = the participation rate.

With known participation rates and a population forecast, it is possible to forecast the number of participants. For example, if $r = 0.05$ for football, and $P = 100\ 000$, then

$$N = rP \qquad \qquad (7.2)$$
$$N = 0.05 \times 100\ 000 \qquad (7.3)$$
$$= 5000$$

An elaboration of this basic method uses age/sex specific participation rates applied to a cohort survival population forecast (see Chapter 2). An example is shown in Table 7.3.

Table 7.3   *Cohort participation method*

Age group (males)	Participation rate (1)	Population projection (2)	Participants (3) = (1) × (2)
5–14	0.10	10 000	1 000
15–19	0.15	8 000	1 200
20–24	0.10	7 500	750
30–44	0.05	8 000	400
			3 350

Note: Assuming $r = 0$ for males <5 and >45 and for all females.

It is then necessary to convert the number of participants into a requirement for facilities. In order to do so it is necessary to estimate the level of participation, that is, the intensity of use made by an individual participant. For example, it has been estimated (Sports Council, 1968) that each football player plays, on average, once a week and that each pitch is used twice a week. Thus, there will be 44 participants per pitch per week, and so from Table 7.3

$$\text{Number of pitches required} = \frac{3350}{44} = 76 \qquad (7.4)$$

(assuming each player plays for only one team).

This simple calculation can be adapted to most activities and can be used to find out the requirements for facilities in a newly planned area or to identify current or futures gaps in provision. It is, however, not without problems:

1   The fundamental problem with the technique is the calculation of participation rates. Typically these are derived from national surveys, are assumed constant, and account may be taken of geographical or social class differences. However, the use of 'standard' participation rates can turn what appears to be a forecast into a more elaborate

version of normative planning (see Section 7.3). Moreover, it is very difficult to forecast participation rates. For well-established sports such as football it may be reasonable to assume *r* to be constant, but for expanding sports such as squash a forecast is necessary. The simplest way to do this is to undertake a trend-line projection (see Chapter 2 and the section on trend lines below).

2   As with all such recreation forecasts a problem is the relationship between supply and participation, as participation may be *supply constrained* (see Section 7.2). Thus additional provision would allow latent demand to be realized. It might also divert demand. For well-established activities such as football this is less likely to be a problem. There may also be substitution — people who turn to squash may then play less tennis or badminton, with consequences for other facilities.

3   Changes in management may accommodate rising demand by, for example, longer opening hours, encouraging non-peak usage, improving the quality of facilities (for example, floodlights), etc. However, the effect may be to remove a supply constraint (as in (2) above).

4   Participation rates are influenced by a large number of factors other than age and sex, such as income, working hours, levels of education, and mobility. Some research has shown that social groups do not provide a reliable basis for forecasts of participation (see Field and O'Leary (1973)), although Snepenger and Crompton (1984) have refined the technique with some success. Rather than produce highly disaggregated rates, multiple regression has been used (see Section 7.4.3).

7.4.2 Trend lines
Trend lines have been fully discussed in the context of population forecasts in Chapter 2. As outlined there, the basic assumption is that the forecast dependent variable will follow a predictable pattern, and thus that time is the independent variable. Trend lines may be used for numbers participating or participation rates. Such a technique is simple and for recreation may have some merit for periods of a few years. For longer periods it is unlikely to be reliable. Moreover, information on participation is frequently unavailable. Information on money spent on sports is more readily available but it is difficult to convert this into numbers of participants. Accordingly the use of the technique is limited.

7.4.3 Multiple regression
Multiple regression has already been considered in Chapters 2 and 3 and in Appendix A.9 and so is not discussed here in depth. In recreation studies it has been used to model participation rates. The model may then be applied

to the same area for some future date or to a different area. The participation rate is the *dependent* variable, while the *independent* variables might include: proportion of population in particular age groups; socio-economic structure; marital status; family structure; household size; leisure time; car ownership, etc. The selection of appropriate explanatory independent variables is not a rigorous process. These will depend not only on the activity being considered but also on data availability. It is, therefore, difficult to generalize. Table 7.4 shows the factors used in one particular study. This is one variant of the regression technique (for full details see North West Sports Council, 1972). In this case the dependent variable was the probability (P) of an individual's participating. An elaboration of this method was

Table 7.4 *Multiple regression and participation*

Dependent variables	*Factors*
Participation rates for: • taking of day trips • taking of half-day trips • turf sports • indoor dry sports • outdoor water sports • rural area sports • swimming • fishing • golf • tennis • bowls	• stratum of residence • sex • age • socio-economic group • marital status • number of children aged under five in the household • number of children aged from 5 to 9 inclusive in the household • number of children aged 10 to 14 inclusive in the household • total number of persons in the household • estimated number of hours worked in an average week excluding weekend working • estimated number of hours worked in an average weekend • estimated number of hours spent in chores in an average week (including weekend) • estimated number of hours of leisure in an average week (including weekend) • possession of a driving licence for a car • ownership of a car

Source: North West Sports Council (1972).

produced by Cicchetti *et al.* (1969) who introduced a supply variable (*S*) in addition to demand variables (D_i). Thus:

$$P = a + \sum_{i=1}^{n} b_i D_i + cS \qquad (7.5)$$

where $a, b_1, \ldots b_n, c$ are constants. (For a fuller discussion see Gratton and Taylor, 1985).

As previously discussed multiple regression has many limitations (see Chapters 2 and 3 and Appendix A.9). In the context of recreation forecasts the most problematic is the assumption that the relationship between dependent and independent variables is constant over the period of the forecast. This is unlikely for many types of recreation activity. Moreover, many of the variables may be intercorrelated (Clawson and Knetsch, 1966; Wilkinson, 1973). Anyone intending to apply regression to recreation forecasts is advised to refer to the studies mentioned above.

7.4.4 The Delphi method
In ancient Greece, Delphi was the home of the oracle — an infallible source of knowledge. In decision-making and forecasting the Delphi method is the name given to a technique which relies on the informed and considered views of experts, rather than on a more formalized numerical approach. The actual structure of the technique varies depending on the circumstances and the topic, but the basic method proceeds in the following stages (Gratton and Taylor, 1985);

1 A large sample of experts is asked, for example, to ascribe probabilities to a number of events, or to assess the total number of participants in an activity. This is done individually and independently.
2 The views of the experts are then collected and collated to produce a frequency distribution.
3 The frequency distribution (and perhaps brief outlines of individual reasoning) is distributed to the experts.
4 The experts are then asked to submit revised assessments, or to state their reasons for not changing their views — particularly if it is not a widely shared view.
5 The process may then be repeated.

In this way it is hoped that a clear consensus view will appear. Face-to-face contact is not necessary, and indeed may be counterproductive if it allows the views of dominant personalities to prevail.

Table 7.5 and Fig. 7.1 show an example of a hypothetical simple three-round application of the method. In this case a consensus emerges quickly,

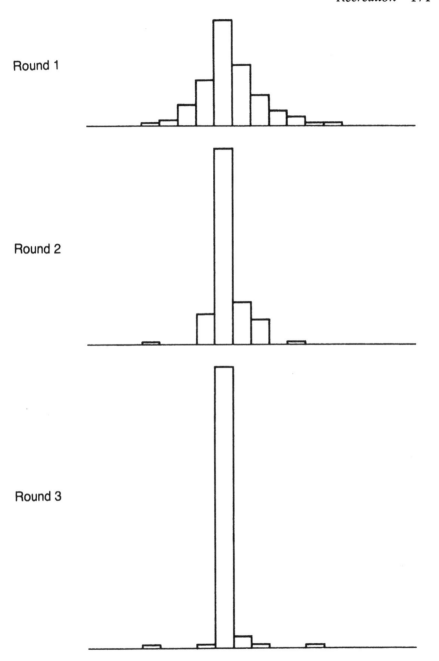

Figure 7.1 *The Delphi method.*

Table 7.5   *The Delphi method*

Question: What will be the average weekly attendance of Football League matches in 2001?

Attendance ('000s)	Number of experts		
	Round 1	Round 2	Round 3
301–325	1	1	1
326–350	2	—	—
351–375	7	—	—
376–400	15	10	1
401–425	35	65	93
426–450	20	14	3
451–475	10	8	1
476–500	5	1	—
501–525	3	—	—
526–550	1	1	1
551–575	1	—	—
	100	100	100

although some of the experts hold their ground and remain outside the majority. Clearly, the basic method can be elaborated by adding questions on dates of events, their likely impact, the desirability of certain outcomes, etc.

This technique, as suggested by Gratton and Taylor (1985), is of unknown and largely untested value in this context. The requirement for a large number of experts makes its use, at other than the national level, somewhat unlikely. (It could be applied to other topics covered in this book; however, for these, established techniques make it unattractive.)

7.5 Forecasting for sites or areas

The methods used to forecast recreation use for particular sites or areas vary somewhat from those outlined above for activities. They fall into three categories, the first of which should by now be very familiar:

1 trend lines
2 the Clawson method
3 gravity models.

For most sites or areas, capacity is an important consideration.

7.5.1 Trend lines

Trend lines are rather more useful for sites or areas than for activities, mainly because attendance data is likely to be more readily available. Information could be collected on an annual or a monthly basis. In most cases simple straight-line trends would be used unless the data clearly indicated that some alternative was more appropriate. For sites with a number of activities, different user types, day and overnight visitors etc., visitor numbers may be disaggregated and separate trend lines produced. Two particular cases are worthy of further consideration:

(a) capacity
(b) seasonality.

(a) Capacity

Any site or resort has a finite upper limit to its capacity to accommodate recreationists. Such capacities could be increased by improving or increasing facilities or by extending opening hours. Indeed, such action could be in response to a forecast of above-capacity demand. Under such circumstances it has been suggested (Butler, 1974) that a logistic curve (see Chapter 2) best represents the pattern of growth (Fig. 7.2). There is an initial period of slow growth followed by a rapid growth rate and then by a much reduced rate as capacity is reached and people find alternatives. In particular this model has been used to describe the growth of tourist resorts.

While this pattern may describe the growth of resorts it is unlikely to have much merit for smaller sites. For such sites the period of development to capacity may be brief or may have been achieved, and so interest lies in trends thereafter. For these sites, management decisions may develop capacity or attract visitors by advertising. As the method describes a pattern of growth rather than explains causal relationships it may be of little value. For

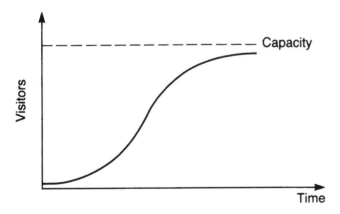

Figure 7.2 *The logistic curve.*

larger resorts the model is of limited use when they begin to decline as changing preferences take visitors elsewhere. As with smaller sites a further problem is the relationship between capacity and provision. As resorts develop so too do facilities, which in turn influence visitor numbers. In general this method ignores the complexity of causal factors.

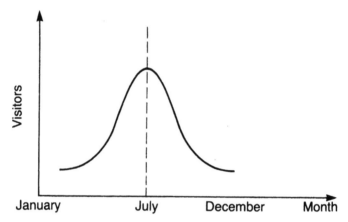

Figure 7.3   *Seasonal trends.*

(b)   Seasonality

Many sites, particularly those associated with outdoor rural activities, have a marked seasonal trend in visitors or users. Typically the seasonal trend is as in Fig. 7.3. However, this seasonality may conceal an underlying trend as shown in Fig. 7.4. In this case the true trend is of linear increase. More complex variants involve a cyclical component to the true trend or non-linear trends. These are not considered here. The method outlined below involves calculating an average seasonal variation and eliminating this from

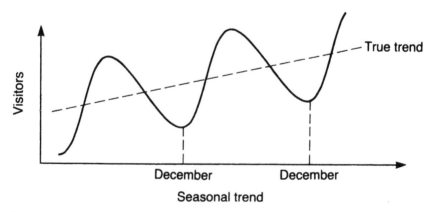

Figure 7.4   *Seasonality and the true trend.*

the data to produce a linear trend line. From this, forecasts may be produced and the seasonal adjustment reapplied to forecast future figures with a seasonal component.

A worked example is shown in Table 7.6 (overleaf). The steps involved are as follows:

(a) From the unadjusted visitor numbers (column (1)) obtain a yearly figure (column (2)).
(b) From (2) obtain a weighted sum of five quarters (3).
(c) From (3) obtain the moving average (4).
(d) The total fluctuation is then (5) = (1) − (4).
(e) The seasonal fluctuation is obtained by reordering column (5) in the second part of Table 7.6, to obtain average seasonal fluctuation (6).
(f) The seasonally adjusted figure is obtained by subtracting the seasonal fluctuation from the unadjusted figures ((7) = (1) − (6)).
(g) The residual fluctuation is obtained by subtracting the moving average from the seasonally adjusted figures ((8) = (7) − (4)).

By this method the actual figures (A) are broken down into three components: a trend line (the moving average) (T); a seasonal fluctuation (S); and a residual fluctuation (R). Thus:

$$A = T + S + R \tag{7.6}$$

The residual fluctuation component (R) is the product of a variety of random factors, and so a forecast can be based on the trend line and seasonal components. The trend line may be determined by any of the methods described in Chapter 2 and an appropriate seasonal adjustment then made. Readers should now forecast the visitor figure for the third quarter of 1988.

This technique is prone to all of the problems previously described for trend-line forecasts. In addition, the assumption of constant levels of facilities may not hold. A particular problem may be that the method ignores the cost of using the facility, which is considered in the following sections. Nonetheless it can be of some value for established facilities with significant seasonal variations.

7.5.2 The Clawson method
This is a method which explicitly considers the cost involved in a recreational visit. It seeks to explain the volume of visits by the cost of making a visit. The effect of changes in costs on visitor numbers may then be forecast. It may also be of value in forecasts of likely usage of proposed new projects based on similar existing sites. The Clawson method has been used widely in academic studies of parks and reservoirs, etc., but has been used rarely in practice.

Table 7.6 *Seasonal adjustment*

Year	Quarter	(1) Unadjusted visitor numbers	(2) Σ4	(3) Σ8	(4) Moving average	(5) Total fluctuation	(6) Seasonal fluctuation	(7) Seasonally adjusted	(8) Residual fluctuation
1981	I	256					−730	986	
	II	992	3584				+134	858	
	III	1984	3648	7232	904	+1080	+1240	744	−160
	IV	352	3872	7520	940	−588	−644	996	+56
1982	I	320	4288	8160	1020	−700	−730	1050	+30
	II	1216	4384	8672	1084	+132	+134	1082	−2
	III	2400	4544	8928	1116	+1284	+1240	1160	+44
	IV	448	4608	9152	1144	−596	−644	1092	−52
1983	I	480	4762	9280	1160	−680	−730	1210	+50
	II	1280	4704	9376	1172	+108	+134	1146	−26
	III	2464	4672	9376	1172	+1292	+1240	1224	+52
	IV	480	4832	9504	1188	−708	−644	1124	−64
1984	I	448	4928	9760	1220	−772	−730	1178	−42
	II	1440	5024	9952	1244	+196	+134	1306	+62
	III	2560	5088	10 112	1264	+1296	+1240	1320	+56
	IV	576	5056	10 144	1268	−692	−644	1220	−48
1985	I	512	5248	10 304	1288	−776	−730	1242	−46
	II	1408	5280	10 528	1316	+92	+134	1274	−42
	III	2752					+1240	1512	
	IV	608					−644	1252	

Table 7.6 *Seasonal adjustment – continued*

| | | Quarter | | | |
		I	II	III	IV
	1981			+1080	−588
	1982	−700	+132	+1284	−596
Year	1983	−680	+108	+1292	−708
	1984	−772	+196	+1296	−692
	1985	−776	+92		
Total fluctuation		−2928	+528	+4952	−2584
Average fluctuation		−732	+132	+1238	−646 $\Sigma = -8$
Adjustment		+2	+2	+2	+2 (to ensure sum of annual fluctuation is zero)
Seasonal constants		−730	+134	+1240	−644

The method, in summary, requires the calculation of a demand curve. Information is collected, by survey, on visitor numbers during a given time period and on the origin of their trips. Visitor rates are calculated for a number of zones and the relationship between this rate and the cost of the visit is determined. The basic structure of the method is outlined in the simple example below.

Table 7.7   *The Clawson method*

Zone	Distance (miles)	Travel cost (at 20p/mile) (£)	Zone population ('000s)	Visitor rate (per 1000 population)	Number of visitors
A	5	1	500	0.4	200
B	10	2	600	0.3	180
C	15	3	400	0.2	80
D	20	4	300	0.1	30
E	25	5	200	0.05	10
					500

Note: for a cost of greater than £5, visitor rate = 0.

Table 7.7 gives the relevant information for a hypothetical site with five zones of origin. This information is graphed in Fig. 7.5 to produce the *initial demand curve*.

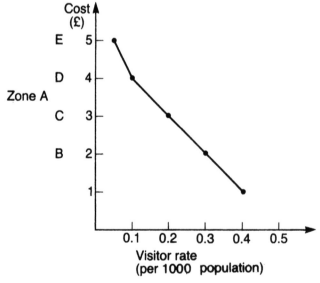

Figure 7.5   *The Clawson method — initial demand curve.*

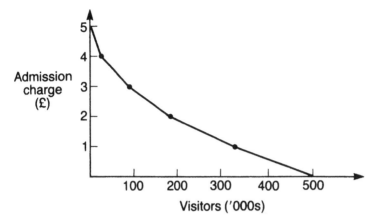

Figure 7.6 *The Clawson method — final demand curve.*

The *final demand curve* is then produced by considering the effect of an admission fee on visitor numbers. At a cost of zero there are 500 visitors. If an admission charge of £1 were made, the total cost for zone A would rise to £2 and so the visitor rate would fall to 0.3, giving 150 visitors. Similarly, visitor numbers from zones B, C, D, E would be 120, 40, 15, 0, giving a total of 325. (Readers should now derive the visitor figures for admission prices of £2, £3, £4, £5.) This information may then be graphed to produce the final demand curve shown in Fig. 7.6. Consumer response to changes in admission price may then be assessed.

There are, however, a number of serious problems associated with this technique (see Mansfield, 1971; R. J. Smith, 1971; Vickerman, 1974).

1   Estimating travel costs is not straightforward (see Chapter 6). Travel could be the cost of petrol or, in addition, could include general running costs of the car and an element for depreciation. Further, as zones of origin are used it is necessary to calculate an average distance from that zone (see Chapter 5). Different numbers of visitors in each car produce differences in individual costs. To overcome some of these problems it is possible to assess motorists' perceived costs. However, as R. J. Smith (1971) has shown, different estimates of travel costs produce substantially different demand curves.

2   A further complication is that visitors tend to perceive entrance costs as more important than travel costs and so adding these to produce the demand curve may lead to problems.

3   Even if the above problems can be overcome and points can be identified through which the demand curve is to be fitted, R. J. Smith (1971) has shown that the *shape* of curve fitted can significantly affect demand forecasts.

4   Traditionally, in transport modelling, the time taken for a journey is regarded as a cost (see Chapter 6). However, in recreation the journey itself may be part of the 'total recreation experience' and so have a benefit rather than a cost.

5   Demand is not homogeneous: it has a number of different components depending, for example, on the length of a particular visit. Holidaymakers are very unlikely to be affected by travel cost from their place of residence to a site.

6   The method assumes that visitors travel by car. Other modes of transport pose problems which, given the limitations of the technique, are unlikely to be overcome by disaggregation.

7   The central assumption of the model — that visitor rate is determined by cost alone — is seriously flawed. Factors such as income, age, car ownership, etc. all influence visitor rates. To overcome this problem some studies have divided the study area into zones of approximately similar socio-economic characteristics. Zoning may then become a very *ad hoc* procedure, particularly if there are no obvious groupings of population centres. In any case, zoning blurs the distinguishing features of the zone by resorting to average characteristics, yet the purpose in producing 'similar' zones is to take into account other demand factors which are assumed to operate at the level of the individual.

8   Further problems arise from ignoring competing facilities which may be of importance in determining visitor rates.

The central problem in recreation forecasting of the relationship between supply and consumption also applies to this method. Finally, there is the ever present problem in forecasting of assuming that the relationships between independent and dependent variables are constant. Given such substantial problems the Clawson method has limited practical use. It should be used with extreme caution and only in its simplest form.

7.5.3 Gravity models

Gravity models have been discussed in detail in Chapters 5 and 6 and so discussion here is limited. The basic equation of a simple gravity model is shown in (7.7) below.

$$V_{ij} = \frac{G P_i A_j}{D_{ij}^b} \qquad (7.7)$$

where V_{ij} = the number of visitors from zone i to facility j
   P_i = the population of zone i
   A_j = the attractiveness of facility j

D_{ij} = the distance between zone i and facility j

G, b = constants.

Once a gravity model framework has been established, the impact of a new development may be assessed in much the same way as for shopping (see Chapter 6). Its estimated level of use and its impact on other facilities can be calculated. Alternative locations may be tested. Information can be calculated on likely traffic generation. However, as with all recreation forecasting, factors such as latent demand (see Section 7.2) cannot be taken into account.

Variants of the gravity model have been used with some success to describe travel patterns to isolated parks and fishing lakes from a number of cities (Boyd-Wennergren and Nielson, 1970; Cheung, 1972) in North America. Application to a densely populated country such as Britain is much more problematic.

Several problems arise with this method:

1 It is necessary to have a closed system of origins and destinations which requires prior study of travel patterns.
2 Zone 'properties' such as distance to a facility are assigned to a single point. If the zones are large this may not be a reasonable assumption (North Londoners are less likely than South Londoners to visit Kent for a day trip because of the necessity to travel across London and lesser knowledge of facilities).
3 Defining recreational attractiveness is very difficult as it may depend as much on perceptions and knowledge as on actual facilities. Some studies have used the simple measure of surface area, but for more elaborate facilities this is inadequate.
4 Distance has a varying influence and so a more complicated distance decay factor may be required. A factor such as D^{-b} overestimates demand for short trips and underestimates demand for longer trips (see Section 5.3.2). As with the Clawson method travel itself may have a positive value.
5 Different types of trips are influenced in different ways by distance. Weekend visits are less influenced. In addition, while the proportion of households from a zone using a facility increases as distance decreases, so the numbers of trips from each household also increases. Distance, therefore, has a dual effect on visitor rates.
6 Factors other than distance affect visitor numbers and distance itself acts differently on different groups. Disaggregation is complex and unlikely to be productive. Car ownership may be taken as a simple surrogate socio-economic variable.

7.6 Conclusions

Increasing recreation activity has made the need to plan more important, particularly as there is substantial government involvement in provision. In this context forecasting should provide the basis for government initiatives and for policies to direct the private-sector. However, public-sector aims of efficiency and equity of provision may conflict with private-sector objectives of profit maximization. Furthermore, the large variety of types of recreation, each with different roles for public and private sectors, makes recreation planning particularly difficult.

These problems are compounded by the inadequacy of forecasting techniques which vary in type, reliability and utility from recreation activity to activity. There is no standard method. In general the forecasts are likely to be unreliable and to be limited in applicability by data availability. One of the most important limitations to recreation forecasting is the relationship between supply and consumption. In many instances consumption is supply led, and so policy decisions which affect the supply of facilities affect future consumption levels. Thus, forecasting is particularly difficult. Levels of consumption do not represent economic demand for they are supply constrained and there may be no cost of consumption.

A further problem is the 'future like the present' assumption of most techniques. It is assumed that relationships between, for example, levels of income and participation rates will be constant. This is typically quite inadequate for recreation, yet without 'clairvoyancy' there seems to be no real alternative. In general the complex causal structures involved in recreation are poorly understood. Even some of the more appealing assumptions may be seriously flawed. Car ownership is typically regarded as a cause of recreation consumption, yet it may be an effect. Cars have an important recreation function and so may be acquired to facilitate participation. Throughout recreation forecasting there is also an assumption that middle-class recreation values will be transferred to other classes as incomes levels rise.

Forecasting and planning should be linked but, with unreliable models whose record of accuracy is poor, it is not surprising that this is rarely the case. At the local scale, where the models are more difficult to use and where data limitations may be more severe, there is often recourse to planning based on standards. The limited forecasting that does take place may be used to validate a policy, to choose between plans, or to assess the impact of a proposal. When, as is often the case, supply is constrained, this may not be a problem.

For most local authorities, a focus on key issues or activities for the resident population is likely to be a reasonable starting point. Participation rates and an inventory of facilities would form a simple basis for action. Comprehensive approaches are unlikely to be efficient or effective. For authorities with substantial numbers of visitors a different approach is necessary, and trend lines seem a reasonable starting point.

These concluding remarks may seem evasive, but generalization is difficult in recreation forecasting. The limited role of planners and the involvement of specialist agencies make the use of simple techniques the most sensible approach for most authorities.

8 Integrated forecasting

8.1 Introduction

Previous chapters have dealt with topic-specific forecasts. However, it should be obvious that there are important interrelationships between these forecasts. Central to the planning of an area are the forecasts of population, housing and employment. Not only do these key topics provide the background to the other topics covered by this book — shopping, transport, and recreation — but they are closely interlinked in a circular scheme of causal relationships. The level and structure of population are important factors in the need or ability to attract employment and the requirements for housing. On the other hand, the growth of employment opportunities may be the main cause of in-migration, or the lack of employment may result in out-migration. Furthermore, lack of available housing may act as a constraint on in-migration to fill vacant jobs, or new house building may result in in-migration, either of these who work elsewhere or associated with employment growth. These interlinkages have led to the development of *integrated forecasting systems*. In this context the integration refers to the use of a set of common, compatible supply-and-demand assumptions for population, housing and employment, with explicit consideration of their relationships. This may be termed *horizontal integration*. There are, however, other aspects to integration, methodologies for which are little developed. These are (Cockhead and Masters, 1984):

1 *Vertical integration:* to ensure the compatibility of forecasts for large areas and long terms with those for smaller areas and shorter terms.
2 *Organizational integration:* these are not based on techniques but on ensuring that different parts of an organization use consistent forecasts.

(These are discussed more fully in Section 8.4.)

There is no standard method of horizontal integration of forecasts. Several versions have been developed by particular local authorities to suit their own needs. Attempts to move to a more standardized approach are not well advanced. Consequently, it is considered appropriate to outline the approach of two particular authorities — Gloucestershire County Council and Grampian Regional Council. First, however, it is necessary to outline briefly the more traditional approach.

8.2 The linear-deductive approach

The traditional approach to forecasting is usually termed *linear-deductive*. In this approach the output of one forecast is used as an input to the other forecasts. All resultant forecasts are assumed feasible and become planning targets; no account is taken of the constraining effect of one forecast level on another. For example, an employment demand forecast would generate a population forecast which would in turn produce a housing 'demand' forecast, and no consideration would be made of the constraining effect of an under-supply of housing. This was the approach of the South Hampshire Structure Plan of 1972. The basic scheme is shown in Fig. 8.1 (overleaf). Employment is seen as the *prime mover* and so this forecast is the starting point from which other forecasts are produced. However, the interrelationships and constraining effects of one forecast on another appear to be ignored: in the language of systems theory, there are no *feedback loops*.

A similar linear-deductive approach was found by Bracken and Hume (1981) in their examination of Welsh structure plans. However, in these plans, population rather than employment was the starting point or prime mover. The population forecast was then used to forecast households and the labour force. The choice of the prime mover depends on the circumstances prevailing in an area. For example, in an area undergoing rapid employment growth, employment would be used. Without such growth pressure it is probable that population would be used. Bracken and Hume (1981) concluded that the linear-deductive approach had some advantages in that it could be quick and cheap to use and, as it was not complex, could be easily understood by planners, politicians and the public. However, there are a number of important limitations:

1 It is *static* — in reality changes are occurring continuously and the model does not reflect the *dynamics* of change.
2 It ignores the mutually constraining *interrelationships* between key activities. The emphasis is on parameters directly related to each activity rather than on the interrelating parameters.
3 It rarely includes the anticipated effects of *policy*. There is a tendency to take forecasts as targets and to assume that activity levels, as

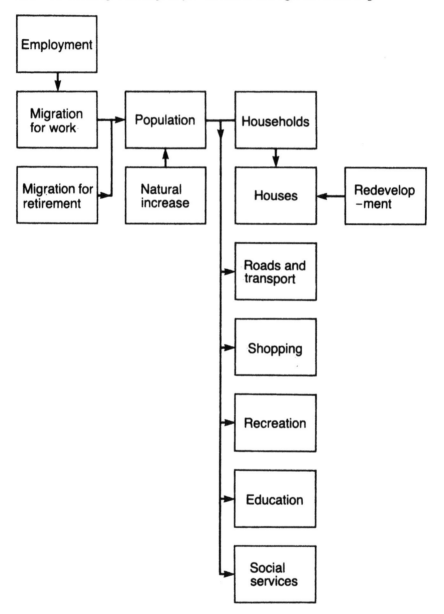

Figure 8.1   *The South Hampshire Structure Plan Approach (Reproduced with kind permission from Breheny and Roberts (1978)).*

opposed to spatial distribution, are not the subjects of policy. In this way forecasts are often made with little regard to planning powers and represent a confusion of wishful thinking and muddled intervention. As such there is no clear guidance for public or private investment.

4 It is usually principally concerned with the supply parameters of population, labour and households, rather than demand.

Integrated forecasting gained impetus according to Breheny and Roberts (1978) from:

1 A dissatisfaction with the standard individual methods with their sequential sets of forecasts which seldom allowed for constraints or feedback.

2 The need to assist experimentation of assumptions and so focus on the more sensitive assumptions.

The principles of integrated forecasting may thus be stated:

1 Separate out integration as a particular and central stage in forecasting.

2 Use a set of common and mutually compatible assumptions.

3 Focus on supply–demand relationships and specify the linkage parameters in their simplest form.

4 Identify the key points where policy measures can exert an influence so as to direct or control outcomes.

Academic work drawing attention to these methodological problems was largely ignored in the first round of structure plans. Since then more attention has been paid to the linkages and constraints. Attempts have been made to link the different forecasts by the parameters of migration, economic activity levels, travel-to-work areas, household formation, etc. As yet no standard methodology has evolved, although individual component forecasts use the standard methods such as cohort survival, headship rates, and shift-and-share analysis as described in previous chapters. The following sections will, therefore, consider individual examples of integrated forecasting.

8.3 The Gloucestershire method

The problems identified above led to the development in Gloucestershire of a basic integrated forecasting model which examined the supply–demand relationships within and between population, employment and housing. The basic structure is shown in Fig. 8.2. Activity levels are related through the supply–demand relationships. Determination of actual levels cannot occur

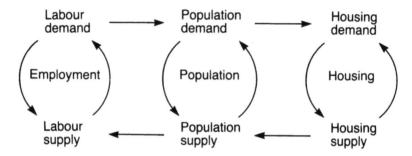

Figure 8.2   *Demand and supply relationships between employment, population and housing (Reproduced with kind permission from Breheny and Roberts (1978)).*

until supply and demand are mutually adjusted. One activity level is held constant and the others are adjusted to produce an equilibrium. In reality, of course, there is no starting point and so it is necessary to 'break in' to the loop and to identify the prime mover activity from which to start. As such it is similar to the linear-deductive approach.

The basic features of this model are (Breheny and Roberts, 1978):

1   The activity levels are determined by the relationships between supply and demand.
2   The supply and demand factors are influenced by, and influence, the levels of other activities.
3   The prime mover in the system is the one placing the strongest constraints on the others — once found, other activities have to be made consistent.
4   The basic activities considered are employment, population, households and housing.
5   The system produced has an equilibrium static approach. It does not model the dynamic and possibly continual imbalance of reality.
6   It is not a model of the causal relationships between component modules, but merely an accounting device to ensure compatibility.

From the outset a distinction is made between outcomes over which policy could exert control and those over which it could not. The procedure is in three stages:

1   Forecasts are made for individual activities without the constraining effects of other activities. This represents the unconstrained potential for change.
2   Forecasts are made with the mutually constraining effects. This produces the integrated forecast.

3   The effects of policy constraints on the integrated forecast are tested.
    These can produce the basis for a forecast which combines planned
    and unplanned changes.

From the basic structure shown in Fig. 8.2 more specific linkages can
be derived. These are shown in Fig. 8.3, which is constructed on the
assumption that:

(a)   Employment tends to be the prime mover which determines general
      demand levels (left to right along the top of Fig. 8.3).
(b)   Housing tends to be the prime determinant of supply levels (right to
      left along the bottom of Fig. 8.3).

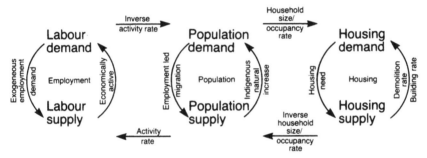

Figure 8.3   *Parameters linking employment, population and housing
(Reproduced with kind permission from Breheny and Roberts (1978)).*

The steps taken are summarized in Fig. 8.4 and are as follows:

1   Forecast the unconstrained employment-labour demand (*UE*).
2   Multiply *UE* by the inverse activity rate (persons per job) and adjust
    for journey-to-work over study area boundary. This produces a
    constrained forecast of resident population (*CP*). (It may be used to
    forecast migration.)

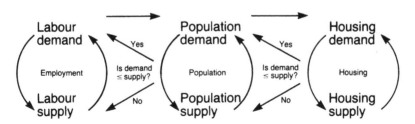

Figure 8.4   *Basic postulated structure of an integrated forecasting system
(Reproduced with kind permission from Breheny and Roberts (1978)).*

3. Apply headship rates to the population to produce a constrained housing demand forecast (*CHD*) (note this is *not* economic demand — see Chapter 3).

1–3 above produce an initial set of demand-based forecasts.

4. Forecast the housing supply (as explained in Chapter 3) to establish unconstrained housing supply (*UHS*).
5. Compare housing supply and demand forecasts. If supply is greater than demand, then there are no housing constraints and calculation stops.
6. If housing demand is greater than supply, assumptions concerning sharing and vacancy rates can be adjusted to compensate. If, after this, demand still exceeds supply, it is necessary to recalculate demand to ensure compatibility with the supply forecast.
7. From the reduced housing figure a reduced population figure must be calculated (*CP*).
8. From the reduced population figure a labour supply figure must be calculated. Adjustments may be made to assumptions on activity levels, unemployment, journey-to-work patterns etc. If demand still exceeds supply, then forecasts are repeated with the labour-supply figure (*CE*).
9. The process is repeated until labour supply and demand match.

From this, separate shorter algorithms were devised for the main cases of use, without prejudicing the original principles (see Breheny and Roberts, 1978, for a full discussion).

In reality adjustments are dynamic and an equilibrium may never be reached. However, to produce a static equilibrium in the model, assumptions must be made if supply and demand do not match, that is:

(a) If and how marginal adjustments might occur to ensure equilibrium.
(b) Which should take primacy if demand and supply are incompatible even after adjustments?

Answers are required when demand and supply for housing (Step 6) and for labour (Step 8) are compared. Only when adjustments have been made beyond a level of tolerance does supply or demand have to take primacy and act as a constraint. If supply exceeds demand no adjustments are necessary, although some could be made to allow for, say, reduced house sharing.

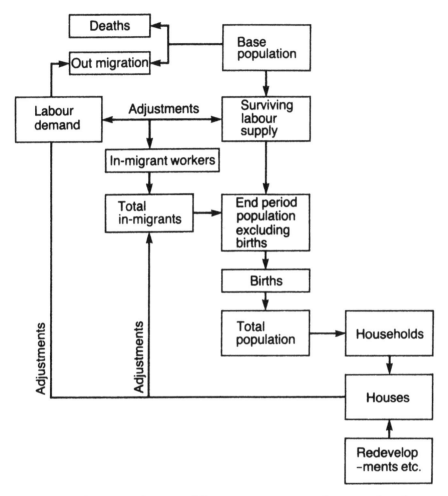

Figure 8.5 *Grampian Integrated Forecasting Framework (Reproduced with kind permission from Grampian Regional Council (1978)).*

8.4 The Grampian methods

During the 1970s the expansion of oil-related activities attracted in-migrants to Grampian Region, but this was constrained by a shortage of housing. In response to these developments and the inadequacies of the linear-deductive approach, Grampian Regional Council (GRC) developed its own model for integrated forecasting (see Fig. 8.5). It was based on three assumptions:

1 Population change consisted of natural increase and migration.

2 In-migration was mainly a function of jobs which could not be filled by the indigenous (after a population forecast based on cohort survival) labour plus commuters.
3 If housebuilding rates did not keep up with forecast in-migration, then a constraint needed to be recognized. Thus feedback was essential.

As can be seen, the general approach is similar to that of Gloucestershire County Council. An attempt was made throughout not to overemphasize the quantitative issues and to include explicit policy assumptions. Individual forecasts used the well-established techniques outlined in Chapters 2–4.

The need to test different policy frameworks and to produce regular updates of forecasts as circumstances changed led to considerable development of integrated forecasting techniques. An important feature was the need to ensure *organizational integration*, i.e. that all organizations with responsibility for the planning of the region used the same forecasts. Functions included in Grampian were education, roads, public transport, water and drainage, police, fire brigade, social work, electoral registration and industrial promotion, as well as land-use planning. Each has differing requirements of time scale and area detail in forecasts.

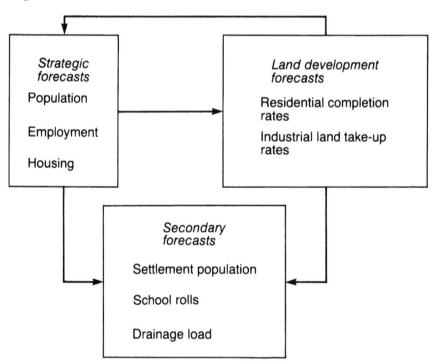

Figure 8.6 *Grampian forecasting framework — the component parts (Reproduced with kind permission from Cockhead and Masters (1984)).*

The framework which was produced for forecasting within the region comprises three major parts (see Fig. 8.6). These are:

1 *Strategic:* integrated forecasts of population, employment and housing for District Council or Structure Plan areas for up to 15–20 years.
2 *Land development:* forecasts of the residential and industrial development rates by individual site for 5–10 years. These were based on current public and private sector intentions.
3 *Secondary:* forecasts of population for small areas to meet the particular demands of service departments.

These forecasts are done on an *annual* basis as shown in Fig. 8.7. Attention here is focussed on the *strategic* forecasts.

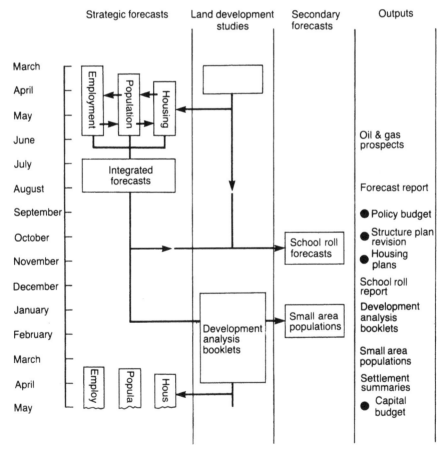

Figure 8.7 *The annual cycle of forecasting in Grampian (Reproduced with kind permission from Cockhead and Masters (1984)).*

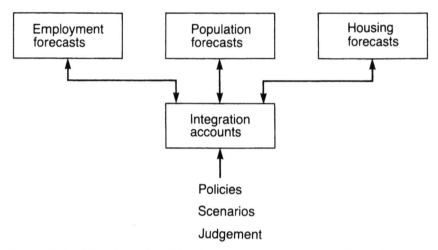

Figure 8.8   *The Grampian federal approach to integration (Reproduced with kind permission from Cockhead and Masters (1984)).*

As demographic change in the region over recent years has been employment led, the employment forecast is central. The oil and non-oil sectors are treated separately (although the division may not always be clear-cut). These are inputs to the population forecasts. Until 1982, GRC used the conventional five-year cohort model. Since then a single-year cohort model has been used as many users required age structures by single years. This is seen as the key to *organizational integration*.

The underlying structure of integration is similar to that of Gloucestershire. Labour demand is assumed to be the key determinant and iteration is used to ensure comparability. As before it is essentially a static accounting framework, albeit with annual updates and more elaborate monitoring of developments. The main linkages are much as before: activity rates, unemployment rates, journey-to-work patterns, headship rates, vacancy and sharing rates, and migration. The process of integration is not rigid and technical but involves judgement and policy input. This is emphasized by Cockhead and Masters (1984), who describe the framework as a 'loose federation' of 'independent activity forecasts'. Despite this emphasis, illustrated in Fig. 8.8, there does not appear to be any substantive difference in technique between the approaches of Grampian and Gloucestershire. What makes the Grampian method distinct is the level of *organizational integration*.

8.5 Integration in Greater Manchester

The former Greater Manchester Council (GMC) developed another variant of integrated forecasting. As with GRC the forecasts were for different user groups with different requirements: County Council planning and service departments; District Council; other public services; and the private sector. A multi-regional approach was developed based on the assumption that in-migration *to* the GMC area was employment based, while movement *within* the area was predominantly for private housing. The model which was developed is too complex to describe in this text, but details can be found in Dewhurst (1984). It is relevant to note that one of the problems which arose was that local authority boundaries did not make the best spatial units for the forecasts, which necessarily involved consideration of housing market areas (see Chapter 3). Such a problem is particularly acute when considering a major conurbation such as the GMC area, but is less of a problem for the Grampian and Gloucestershire areas where settlements are smaller and district council areas are larger.

8.6 Conclusions

Integrated forecasting is an advance on linear-deductive methods of forecasting. However, the techniques outlined in this chapter are integrated only insomuch as the component forecasts are based on a common set of assumptions and are consistent within an accounting framework. Moreover, they are static equilibrium approaches rather than dynamic models. While, in general, integrated forecasts are of more value, individual forecasts are still of use in giving some indication of longer-term changes, for it is inappropriate to forecast activities in equilibrium for other than a few years.

Adjustments made to the linking parameters to ensure equilibrium are without sound theoretical justification. In the past economic activity rates were an alternative response to demand changes. More recently, with recession, factors such as voluntary redundancy, early retirement, short-time working, etc. have further complicated the issue. Changes have also occurred in one-person and single-parent households with a consequent increased demand for housing. In general, the linking parameters have become unstable. It is, therefore, difficult to assess what adjustments might occur.

From the outset it is essential to distinguish between forecasts, usually trend-based, and policy objectives. It is also necessary to be aware which forecast outcomes are controllable and which are not. The scope to influence future outcomes and the flexibility of policy stances should be important concerns of the forecaster. Forecasts and policy factors are inevitably interlinked. For example, Turner (1982) suggests that the compatibility of forecasts has often been destroyed by last minute, dramatic modifications to housing and industrial land allocations without due care to the consequences on other policies based on population forecasts.

But there will always be problems with forecasting. As always an awareness of the limitations leads to more effective use and caution is required. Integrated forecasting has much to commend it and advances in line with developments in Grampian Regional Council seem to offer most scope for improvement and advancement.

9 Conclusion

Planning models and techniques provide systematic statements of relationships between the different elements of spatial structure. Not surprisingly, therefore, an extensive range of analytical methods have evolved in an attempt to understand and quantify the nature of these relationships. It was not the intention of this text to spell out in detail the numerous tools of planning analysis, which range from the basic methods of descriptive statistics to the construction of more complex mathematical models, or, indeed, fully to develop their potential. Rather, the principal objective has been to introduce the reader to the complexities of those techniques in common usage. To this end, the more important existing methods have been presented and their utility in planning analysis has been evaluated. It would be presumptuous to claim that the appraisal is entirely objective or, for that matter, balanced, but it is clearly our view that the analysis of spatial structure, as well as its consequences, is facilitated by the appropriate use of quantitative methods.

Throughout, the text has dealt with those techniques that relate to particular topics or subsystems within the broader planning framework. The approach is not only, by definition, partial, but each technique can also suffer major limitations when employed singly. Meaningful planning analysis is interdependent analysis and it is clear that forecasts should, as a result, be *integrated*, as they are interrelated. For this reason some modest fusion is attempted with the partial techniques in the topic chapters and, subsequently, a wider analytical framework is developed in Chapter 8. It is, nevertheless, realized that this is some way from the kind of comprehensive synthesis whose desirability is constantly advocated by other commentators.

Statistical and mathematical models in themselves are nothing more than abstract representations and simplifications of a particular phenomenon of interest. Because models simplify reality, there is always the danger that they might also end up distorting and obscuring it. An intelligent approach to planning analysis must, therefore, also probe the area of theory, no matter how elegant or mathematically consistent the techniques in question, and even when complemented by an abundance of well-conceived empirical studies. Any critical appraisal of those techniques in current usage needs to acknowledge that they have been developed against a background of relatively limited theoretical knowledge and must not only evaluate the choice of assumptions in the model-building process, but also the mode of abstraction. Experience suggests that there is some justification in Sayer's criticism that

> ... in the case of planning models, the planner is rarely advised to ask whether the model's conceptualisation of the city and urban life is appropriate — the validity of the mode of abstraction is largely taken for granted. All the planner is supposed to worry about are the secondary issues of zoning systems, data sources, calibration etc. . . . (Sayer, 1976, p. 250).

Although techniques can never be substitutes for theory, they can aid understanding and theory formulation. Furthermore, the analytic framework in which they are evaluated frequently leads to better understanding and increasing rigour in the thinking process. The suggestion that their highly technocratic and formalized presentation means that they are often little more than internally consistent sets of equations is certainly an overstatement. This text makes no claim to have considered the theoretical issue in any depth and, indeed, has only touched the surface when discussing theoretical limitations in the topic chapters. But this requires no apologies, since it was the intention to produce a practical introduction to methods of planning analysis.

Associated with the general question of theory there are a number of more specific problems. *Aggregation* is an issue which has been raised in various parts of the book. The tendency in urban modelling is to study change at the micro-level, that is at the level of the household or firm. But there are serious theoretical problems in translating and utilizing new knowledge from research at the micro-level, with the need for more aggregative data consistent with the kind of integrative models discussed in Chapter 8. This is, and is likely to remain, an important limitation in developing more comprehensive planning models.

When using planning models, there is rarely a correct answer. In general, therefore, it is preferable to produce a range of values when making forecasts and to test the sensitivity of these to changes in the underlying assumptions of the model in question. This presupposes a well-defined set of such assumptions, and implies a degree of objectivity in their formulation. Unfortunately, politics and policy are difficult to disentangle from the purely technical aspects of forecasting methodology.

Notwithstanding the above qualifications, the main conclusion to be drawn is that there exists a reasonably reliable standard 'tool kit' of analytical methods that lend themselves to making forecasts in practical planning situations. Such forecasts are necessary, but should not be treated as answers — merely as guides to decision making. Techniques and models must not be allowed to become the be-all and end-all of the planning process but must of necessity remain as very important instruments in the identification, articulation, and solution of spatial problems. Even if planners do not themselves have to use the techniques, an understanding of them and their associated problems and limitations can only lead to better application of the results.

Appendix Basic mathematics

A.1 Introduction

To the non-numerate mathematics may seem like an impenetrable mystery, and so efforts to understand it appear purposeless. This is not the case. Mathematics at the level required for this text is, if not simple, then comprehensible with a little effort. Readers must lower their defences, ignore their instincts, believe the goal is achievable, and concentrate.

The mathematics presented here requires a little thought to come to terms with some basic concepts and then some memory work to learn the language. For the reader willing to make the effort the rewards are enormous: a new world of comprehension will open up.

The concepts to be understood are, typically, intuitively obvious: many are within the experience of all readers, but are expressed, in mathematics, in a more precise, general or abstract form. However, the use of simple numerical examples often clarifies seemingly incomprehensible ideas. Mathematics uses a language, a code, to ensure efficiency and rigour of expression: it is no more than that. With the correct attitude and some effort much can be understood.

Of course, this is not to suggest that the following notes will produce new Einsteins. They are only a collection of simple explanations of relevant concepts. Much fuller explanations may be found elsewhere.

A.2 Operations

For the purposes of this text the number system used is the set of *real* numbers. Precisely what this means is of little relevance here, but in summary it includes positive and negative *integers* (such as $-3, -2, -1, 0, 1, 2, 3$); *rational* numbers (those which can be expressed as the ratio of two integers, such as $^3/_2, ^7/_3, ^1/_5, -^2/_7$); and *irrational* numbers (those which cannot be expressed as the ratio of two integers, such as $\sqrt{3}$ — the square root of 3).

In this number system there are two main *operations* — *addition* (+) and *multiplication* (×). It is hardly necessary to explain what these mean as they are well known to all. This system — the set of real numbers and the operations of addition and multiplication — is only one example of a more general concept. In common with all such systems it obeys a number of basic rules. Thus, taking a, b, c as *any* real numbers, the following always hold:

(a)　The sum of two numbers is the same regardless of the order. In symbols:

$$a + b = b + a \tag{A.1}$$

This is known as the *commutative law*. Using a simple numerical example this law is intuitively obvious:

$$4 + 3 = 3 + 4$$

The commutative law is also true for multiplication, thus

$$a \times b = b \times a \tag{A.2}$$

Or to take an example

$$4 \times 3 = 3 \times 4$$

The multiplication sign is often omitted and $a \times b$ is written as $a.b$ or even ab.

(b)　The sum of a set of numbers is the same regardless of how they are added together. In symbols:

$$(a + b) + c = a + (b + c) \tag{A.3}$$

This is known as the *associative law*. Taking a simple example:

$$(2 + 3) + 5 = 2 + (3 + 5)$$

From the left-hand side (lhs):

$$(2 + 3) + 5 = 5 + 5 = 10$$

From the right-hand side (rhs):

$$2 + (3 + 5) = 2 + 8 = 10$$

The associative law also holds for multiplication, thus

$$(a \times b) \times c = (a \times b) \times c \tag{A.4}$$

Readers should now use a simple numerical example to prove this.

(c)　The sum of two numbers multiplied by a third number is the same as that obtained by multiplying each of the two numbers by the third separately and then adding the two results together. In symbols:

$$a \times (b + c) = (a \times b) + (a \times c) \tag{A.5}$$

This is known as the *distributive law*. Again it is obvious with a simple numerical example:

$$5 \times (6 + 4) = (5 \times 6) + (5 \times 4)$$

From the lhs:

$$5 \times (6 + 4) = 5 \times 10 = 50$$

From the rhs:

$$(5 \times 6) + (5 \times 4) = 30 + 20 = 50$$

When put in terms of simple numerical examples these rules should be familiar to all. Mathematics, in this case, can be viewed as the abstraction of familiar concepts to produce a precise and succinct representation. It may be unfamiliar, but it should not intimidate.

In number systems there are two further operations which are subsidiary to the main two. For real numbers these are known as subtraction and division. Before proceeding to discuss these it is necessary to introduce the concepts of an *identity element* and an *inverse*.

The *identity element* of an operation is the number which, when applied to any number, leaves the number unchanged. For addition the identity element is 0; for example,

$$a + 0 = 0 + a = a \tag{A.6}$$

For multiplication the identity element is 1; for example,

$$a \times 1 = 1 \times a = a \tag{A.7}$$

The *inverse* of a number is the one which when applied to the number through the operation gives the identity element for that operation. Thus

$$a + (-a) = (-a) + a = 0 \tag{A.8}$$

so $-a$ is the *additive inverse* of a.

And:

$$a \times \frac{1}{a} = \frac{1}{a} \times a = 1 \tag{A.9}$$

So $\frac{1}{a}$ is the *multiplicative inverse* of a.

(Note that $a = \frac{a}{1}$ so $\frac{a}{1} \times \frac{1}{a} = \frac{a \times 1}{1 \times a} = \frac{a}{a} = 1$.)

Subtraction may now be regarded as the addition of the additive inverse, so:

$$a - b = a + (-b) \tag{A.10}$$

Division may be regarded as the multiplication by the multiplicative inverse, so:

$$a \div b = a \times \left(\frac{1}{b}\right) \tag{A.11}$$

(Readers should note that, for reasons which need not matter here,

$$a - (-b) = a + b \text{ and } -a \times -b = a \times b)$$

The commutative and associative laws do not apply for subtraction or division, and although multiplication is distributive over subtraction as well as addition, none of the other three operations is distributive over anything.

Once these basics are understood, the algebra used in this book should become straightforward. Some parts may seem blatantly obvious and simplistic, while others may appear obscure. With a little effort any difficulties encountered should be easily overcome.

A.3 Powers

Raising a number to a *power* means multiplying by itself that number of times. a^b is called '*a* to the power *b*'.

For example $a^4 = a \times a \times a \times a$ (A.12)
and $10^3 = 10 \times 10 \times 10 = 1000.$

This holds for all *real number* values of *b*. Although when the power is not a positive integer $(1, 2, 3, \ldots)$ it is more difficult to imagine, this need not give trouble.

The following are some of the basics of 'powers'.

$$a^m \times a^n = a^{m+n} \tag{A.13}$$

$$a^m \div a^n = a^{m-n} \tag{A.14}$$

$$(a^m)^n = a^{mn} \tag{A.15}$$

$$(a \times b)^n = a^n \times b^n \tag{A.16}$$

$$\left(\frac{a}{b}\right)^n = \frac{a^n}{b^n} \tag{A.17}$$

$$a^{m/n} = (a^{1/n})^m = (\sqrt[n]{a})^m \tag{A.18}$$

$$a^0 = 1 \tag{A.19}$$

$$a^{-n} = \frac{1}{a^n} \tag{A.20}$$

(Note that $a^{1/2}$ is the square root of *a*, as $(a^{1/2})^2 = a^1 = a$
Similarly $a^{1/n}$ is the *n*th root of *a*, as $(a^{1/n})^n = a^1 = a$)

Readers should now confirm results (A.13)–(A.18) using $a = 16, b = 8$, $m = 4, n = 2$.

A.4 Summation

Mathematical notation is liberally sprinkled with letters from the Greek alphabet. The symbol Σ (pronounced sigma) is the Greek equivalent of S and is used as shorthand to represent a *sum*. Thus

$$\sum_{i=1}^{5} 2^i$$

denotes the sum from $i = 1$ to $i = 5$ of 2^i. The symbol i is called the index number. In longhand the sum is

$$\sum_{i=1}^{5} 2^i = 2^1 + 2^2 + 2^3 + 2^4 + 2^5 \tag{A.21}$$

More generally $\sum_{i=1}^{n} 2^i$ is *the sum to n terms*. It is often written in shorthand as $\sum_i 2^i$ or $\Sigma 2^i$.

A general series is written $\sum_{i=1}^{n} a_i$, where the general term a_i is unspecified. In longhand:

$$\sum_{i=1}^{n} a_i = a_1 + a_2 + a_3 + \cdots + a_n \tag{A.22}$$

Similarly:

$$\sum_{i=1}^{n} a_i^2 = a_1^2 + a_2^2 + a_3^2 + \cdots + a_n^2 \tag{A.23}$$

The following are some of the rules for algebraic manipulation of series.

$$\sum_{i=1}^{n} (a_i + b_i) = \sum_{i=1}^{n} a_i + \sum_{i=1}^{n} b_i \tag{A.24}$$

$$\sum_{i=1}^{n} ka_i = k \sum_{i=1}^{n} a_i, \qquad \text{where } k \text{ is a constant} \tag{A.25}$$

$$\sum_{i=1}^{n} k(a_i + b_i) = k \sum_{i=1}^{n} a_i + k \sum_{i=1}^{n} b_i, \text{where } k \text{ is a constant} \tag{A.26}$$

(A.24) above is easily proved:

$$\sum_{i=1}^{n} (a_i + b_i) = (a_1 + b_1) + (a_2 + b_2) + (a_3 + b_3) + \cdots + (a_n + b_n) \tag{A.27}$$

Rearranging the terms by applying the *associative law* (see Appendix A.2 above) gives:

$$\sum_{i=1}^{n} (a_i + b_i) = (a_1 + a_2 + a_3 + \cdots + a_n)$$

$$+ (b_1 + b_2 + b_3 + \cdots + b_n) \tag{A.28}$$

$$= \sum_{i=1}^{n} a_i + \sum_{i=1}^{n} b_i \tag{A.29}$$

Readers should now try to prove (A.25) and (A.26) above.

It is irrelevant which letter is used as the index. It is merely conventional to use i — it is short for *i*ndex. When it is necessary to distinguish two summations the letters j, k, l are usually used. Similarly it is conventional to use n for the *n*umber of terms in the series. To distinguish between two summations which may have different numbers of terms, the letter m is used.

Double summations

In mathematics it is common to use what is known as a double *summation*: in shorthand,

$$\sum_{j=1}^{m} \sum_{i=1}^{n} a_{ij} \tag{A.30}$$

At first this may seem impenetrable, but it really is quite simple to understand. Suppose there is an array of numbers. These can be depicted as shown in Table A.1 below:

Table A.1   *An array of numbers for summation*

a_{11}	a_{12}	a_{13}	$\cdots$	a_{1m}
a_{21}	a_{22}	a_{23}	$\cdots$	a_{2m}
a_{31}	a_{32}	a_{33}	$\cdots$	a_{3m}
.	.	.		.
.	.	.		.
.	.	.		.
a_{n1}	a_{n2}	a_{n3}	$\cdots$	a_{nm}

The sum of the first column is $A_1 = \sum_{i=1}^{n} a_{i1}$ $\tag{A.31}$

The sum of the second is $A_2 = \sum_{i=1}^{n} a_{i2}$ $\tag{A.32}$

The sum of the jth column is $A_j = \displaystyle\sum_{i=1}^{n} a_{ij}$ (A.33)

The total of all terms is the sum of the column totals or

$$S = \sum_{j=1}^{m} A_j$$ (A.34)

$$= \sum_{j=1}^{m} \sum_{i=1}^{n} a_{ij}$$ (A.35)

The double summation for all its apparent complexity of symbols is, in essence, a very simple concept.

A.5 Functions

A *function* is a relationship between one set of numbers and another set. If x represents members of one set and y represents members of the other, then x and y are called *variables*. That is, their values vary as they may represent any member of the set.

The statement of $y = f(x)$ (A.36)

means 'y is a function of x', or,
for any value of x, a corresponding value for y can be calculated. In this case it is conventional to call x the *independent* variable and y the *dependent* variable, as y is calculated from x. As always the use of a simple example should clarify. If:

$$y = 3x$$ (A.37)

then the function is 'multiply by 3'. Thus for any value of x the corresponding value of y can be calculated as shown in Table A.2:

Table A.2   *The function $y = 3x$*

x	1	2	3	. . .	—
y	3	6	9	. . .	

Of course, not all functions are as simple as this, but the basic principles are the same.

When dealing with functions it is conventional to use letters such as f, g, h, to denote functions; x, y, z, to denote variables; and a, b, c, to denote constants.

For example $y = f(x) = ax + b$ (A.38)

Some values for this function with $a = 3$ and $b = 2$ are given in Table A.3.

Table A.3 *The function y = 3x + 2*

x	1	2	3	. . .
y	5	8	11	. . .

Similarly if $y = 2x^2 + 1$,　　　　　　　　　　　　　　　　　　　　(A.39)
then some values are as shown in Table A.4.

Table A.4 *The function y = 2x² + 1*

x	1	2	3	. . .
y	3	9	19	. . .

The above are examples of two-variable functions. It is possible to have more than two variables. For example

$$y = f(x_1, x_2, x_3)$$　　　　　　　　　　　　　　　　　　　　(A.40)

denotes a four-variable function. There are three independent variables, x_1, x_2, x_3, and one dependent variable, y.

An example of such a function is

$$y = x_1 + 2x_2 + 5x_3 + 1$$　　　　　　　　　　　　　　　　　　　(A.41)

This gives $y = 9$ when $x_1 = 1$; $x_2 = 1$; $x_3 = 1$, etc.

A.6　The Cartesian diagram

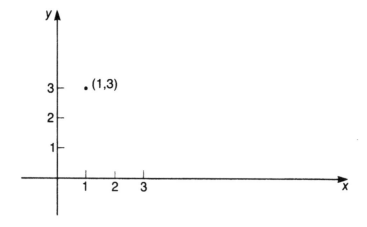

Figure A.1 *The Cartesian diagram.*

Two-variable functions may be graphed on a Cartesian or coordinate diagram as in Fig. A.1. The horizontal axis measures x and the vertical axis measures y. Thus the point $x = 1$, $y = 3$, in shorthand $(1,3)$ is 1 unit 'along' and three 'up'. It is thus possible to graph two-variable functions. Examples are shown in Fig. A.2, with values given in Tables A.5 and A.6.

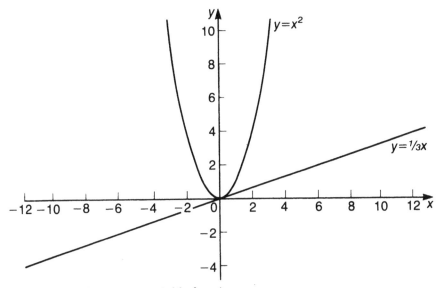

Figure A.2 *Some two-variable functions.*

Table A.5 *A simple linear function*   **Table A.6** *A quadratic function*

$y = \frac{1}{3}x$			$y = x^2$	
x	y		x	y
.	.		.	.
.	.		.	.
.	.		.	.
-3	-1		-3	9
-2	$-\frac{2}{3}$		-2	4
-1	$-\frac{1}{3}$		-1	1
0	0		0	0
1	$\frac{1}{3}$		1	1
2	$\frac{2}{3}$		2	4
3	1		3	9
.	.		.	.
.	.		.	.
.	.		.	.

A.7 Two-variable linear functions

A *linear* or *first degree* function is one with no powers of the variables other than one. Thus

$$y = 3x \tag{A.42}$$

is linear, but

$$y = 3x^2 \tag{A.43}$$

is *quadratic* or *second degree*.

When graphed, a two-variable linear function produces a straight line — hence its name. The general equation for a two-variable linear function is

$$y = mx + c \tag{A.44}$$

where m and c are constants. In the example (A.37) above $y = 3x$, so $m = 3$ and $c = 0$.

m measures the *gradient* of the line. If m is positive, the line slopes from bottom left to top right. The greater is m, the steeper is the line. If m is negative the line slopes from top left to bottom right (see Fig. A.3).

c marks the *intercept* on the y axis. As c inceases, the height of the line increases (see Fig. A.4).

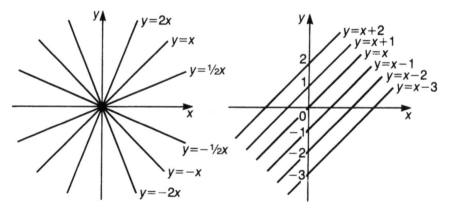

Figure A.3 *The effect of varying m.*   **Figure A.4** *The effect of varying c.*

If any two points *on the line* are known it is possible to calculate the general equation of the line. An example is shown in Fig. A.5. The gradient m is the rise divided by the distance along, or

$$m = \frac{y_2 - y_1}{x_2 - x_1} = \frac{5 - 3}{2 - 1} = \frac{2}{1} = 2 \tag{A.45}$$

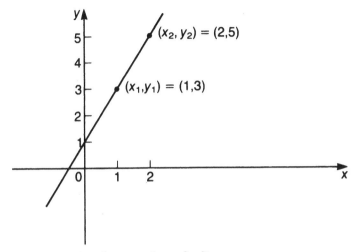

Figure A.5 *Deriving the equation of a line.*

To calculate c, note that for any point on the line $y = 2x + c$, thus for $x = 2$, $y = 5$,

$$y = 2x + c \tag{A.46}$$
$$\text{so } y - 2x = 2x + c - 2x \tag{A.47}$$
$$\text{so } y - 2x = c \tag{A.48}$$
$$\text{and } c = y - 2x \tag{A.49}$$
$$= 5 - (2 \times 2)$$
$$= 5 - 4$$
$$= 1$$

Thus the line is $y = 2x + 1$ \hfill (A.50)

A.8 Linear regression

It is frequently the case that it is necessary to calculate the functional relationship between two variables. That is, given values of x and y, it is necessary to calculate f such that $y = f(x)$, where y is the dependent variable and x is the independent variable.

It is common that the function is linear or approximately linear. The most commonly used method of finding the equation of the line which 'best fits' the data is called *linear regression*. It is possible that some, or indeed all, of the points may not be on this *regression line* (see Fig. A.6).

For each value of x_i, there is a corresponding $\hat{y}_i$ on the line, such that $\hat{y}_i = mx_i + c$. Often $\hat{y}_i \neq y_i$ (see Fig. A.6). This difference $(y_i - \hat{y}_i)$ is called u_i. Thus

$$y_i = mx_i + c + u_i \tag{A.51}$$

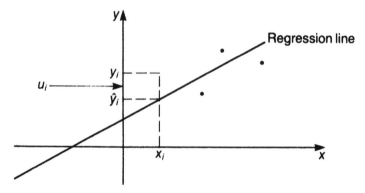

Figure A.6   *The regression line.*

The method of *least squares regression* finds the line $\hat{y} = mx + c$ so as to minimize

$$\sum_{i=1}^{n} (y_i - \hat{y}_i)^2 = \sum_{i=1}^{n} u_i^2$$

u_i is known as the *error term* and $\sum_{i=1}^{n} u_i^2$ is known as the residual *sum of squares*.

The details of the mathematics and the assumptions required need not be dealt with here. An example (Table A.7) should clarify the procedure for calculating the regression line.

Table A.7   *Calculating the regression line*

i	x	y	
1	10.6	72	
2	10.1	77	
3	9.2	89	$n = 10$
4	11.6	62	
5	12.5	55	
6	13.5	46	
7	14.2	32	
8	7.2	93	
9	8.3	91	
10	9.9	81	

The procedure is as follows:

1   Calculate

$$\sum_{i=1}^{n} x_i; \quad \sum_{i=1}^{n} y_i; \quad \sum_{i=1}^{n} x_i^2; \quad \sum_{i=1}^{n} y_i^2; \quad \sum_{i=1}^{n} x_i y_i$$

(Answers: 107.1; 698; 1192·45; 52534; 7068.6)

2 Calculate

$$A = \Sigma x_i^2 - \frac{(\Sigma x_i)^2}{n}; \quad B = \Sigma y_i^2 - \frac{(\Sigma y_i)^2}{n}; \quad C = \Sigma x_i y_i - \frac{(\Sigma x_i)(\Sigma y_i)}{n}$$

(note for shorthand $\overset{n}{\underset{i=1}{\Sigma}}$ becomes Σ)

(Answers: $A = 45.41$; $B = 3813.6$; $C = -406.98$)

3 Calculate:

$$m = \frac{C}{A}; \quad c = \frac{\Sigma y_i}{n} - \frac{C}{A} \frac{\Sigma x_i}{n}$$

(note that $\frac{\Sigma y_i}{n}$ is the mean or average, usually written as $\bar{y}$ and similarly for $\bar{x}$)

(Answers: $m = -8.96$; $c = 165.76$)

4 The regression line is $y = mx + c$.
   (In this case $y = -8.96x + 165.76$.)

It is possible to calculate a regression line for any data set. However, it is usually necessary to determine whether the line is a 'good fit'.

In Fig. A.7 the line is not a 'good fit' as the points are widely scattered and the relationship between x and y is clearly not linear. By contrast Figs. A.8 and A.9 show points which much more nearly fit a straight line.

A simple way to assess if the points have an approximately linear relationship is to calculate the *correlation coefficient*. Using the notation of the regression calculation, the correlation coefficient is defined as:

$$r = \frac{C}{\sqrt{(AB)}} \tag{A.52}$$

The minimum value of r is -1 and the maximum is 1. If r is negative the line

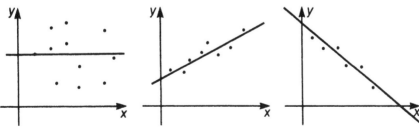

Figure A.7 *Poor correlation.*   **Figure A.8** *Good positive correlation.*   **Figure A.9** *Good negative correlation*

slopes from top left to bottom right. The nearer r is to -1 or 1, then the stronger is the correlation. If $r = 0$ then there is no correlation.

For the regression example of Table A.7,

$$r = \frac{-406.98}{\sqrt{(45.41 \times 3813.6)}}$$

$$= -0.978$$

In this case there is strong negative correlation: as x increases, y decreases.

A.9   Multiple regression

It is possible to extend the ideas outlined above to models with more than one independent variable. Thus:

$$y = a + m_1 x_1 + m_2 x_2 + \cdots + m_n x_n \tag{A.53}$$

or in summary:

$$y = a + \sum_{i=1}^{n} m_i x_i \tag{A.54}$$

In addition to the requirement of a linear relationship between the dependent and independent variables, there is a further requirement of *additivity*. Put simply this merely means that if a new variable is included its effect is added and does not affect other variables. The technicalities of this need not trouble the reader. Those readers wishing to know the details of multiple regression can find full explanations in most books on statistics. Such detail is not required here, but only becomes necessary if the technique is to be used. It is, nonetheless, worthwhile to address a few issues and explain a few terms.

The structure of causality in regression is often misunderstood. Even if a good fit exists between dependent and independent variables, this does not imply a cause-and-effect relationship. For example, it can be shown that the growth in television ownership and the growth in washing machine ownership in the post-war period produce a good-fit regression line. But one does not cause the other — rather both are the effect of rising real income. Care is, therefore, required in drawing conclusions from statistical relationships.

Even if a linear model has been calculated with a sound structure of causality, using it to forecast future values of the dependent variable is problematic. First, there is the problem of forecasting values of the independent variables. Second, there is the necessary assumption that the coefficients (the m_i) are constant. This is a major problem and is referred to throughout the text. Third, the longer is the forecast period then the wider is the range of 'likely' values for the dependent variable. Ideally, *confidence intervals* should be considered, that is a range within which there is a (say)

95% chance of the true value lying. However, the range may be very large. The calculation of such intervals is beyond the scope of this book.

A development of the basic regression technique is to include *qualitative* or *dummy* variables. These are variables which do not have a numerical value but a qualitative value. Examples would be sex — male or female; location — inner city, suburban, rural. It is possible to incorporate such variables into regression analysis, but this is not discussed here.

One common problem with multiple regression is *multicollinearity*. This is present when there is a high level of correlation between the sample observations of independent variables or linear combinations of them. For example, if

$$y = a + m_1x_1 + m_2x_2 \tag{A.55}$$

but $\quad x_1 = 2x_2$, then

$$y = a + 2m_1x_2 + m_2x_2$$
$$= a + (2m_1 + m_2)x_2 \tag{A.56}$$

This is an extreme case used only to illustrate what is a complicated phenomenon. It is clearly necessary to be aware of such relationships, as to include new variables which are linear combinations of other variables adds little to the analysis. There are standard procedures for dealing with this.

A final word of warning is required for multiple regression. In the past there was a tendency to 'throw' data into a computer to see what relationships might emerge. Procedures are now less cavalier, but caution is required in the use of the technique. Anyone considering its use must be well aware of its shortcomings. Some of these are referred to in specific instances in the text.

A.10 Logarithmic and exponential functions

The logarithmic function (log for short) can be explained easily.

If

$$y = a^x \tag{A.57}$$

then $\log_a y = x$ (a is known as the base and x the exponent). In words, the log of y to the base a is x. Or in order to obtain y it is necessary to raise a to the power x.

For example, if the base is 10, $\log_{10} 1000$ is 3, as $10^3 = 1000$.

There are two commonly used bases: 10 and e. The number e is called the exponential constant and is approximately equal to 2.71828. Where e comes from is not of relevance here, suffice it to say that it is widely used in mathematics. (To distinguish from $\log_{10}$, $\log_e$ is often written as ln.)

The *antilog* function is, for the purposes of this discussion, equivalent to an inverse. Thus, generally:

continued overleaf

$$\text{antilog } (\log x) = x \tag{A.58}$$

and

$$\log (\text{antilog } x) = x \tag{A.59}$$

It is useful to think of the antilog of a number as the base raised to the power of that number. Thus

$$\text{antilog}_a x = a^x \tag{A.60}$$

For example, if $x = 3$; $a = 10$; $y = 1000$

$$\log_{10} y = \log_{10} 1000 = \log_{10} 10^3 = 3$$

and $\text{antilog}_{10} x = \text{antilog}_{10} 3 = 10^3 = 1000$

Thus $\text{antilog}_{10} x = y$ and $\log_{10} y = x$

The following are the two main rules for manipulation of logarithms:

$$\log a.b = \log a + \log b \tag{A.61}$$

$$\log a^n = n \log a \tag{A.62}$$

These can be used in algebraic manipulation. Suppose:

$$A = (1 + r)^n B \tag{A.63}$$

$$\text{then } \log A = \log [(1 + r)^n B] \tag{A.64}$$

$$\log A = \log (1 + r)^n + \log B \qquad \text{(from A.61)} \tag{A.65}$$

$$\log A = n \log (1 + r) + \log B \qquad \text{(from A.62)} \tag{A.66}$$

$$(\log A - \log B) = n \log (1 + r) \quad \text{(subtract } \log B \text{ from both sides)} \tag{A.67}$$

$$\frac{\log A - \log B}{n} = \log (1 + r) \qquad \text{(divide both sides by } n) \tag{A.68}$$

$$\text{antilog} \left[\frac{\log A - \log B}{n} \right] = \text{antilog} \left[\log (1 + r) \right] \text{ (take antilogs of both sides)} \tag{A.69}$$

$$\text{antilog} \left[\frac{\log A - \log B}{n} \right] = 1 + r \tag{A.70}$$

$$\text{antilog} \left[\frac{\log A - \log B}{n} \right] - 1 = r \qquad \text{(subtract 1 from both sides)} \tag{A.71}$$

This is the result required in Section 2.2.1.

Further details of logarithms need not trouble the reader here. However, they have one important use in transforming nonlinear equations to linear equations.

A.11 Transforms

It is often the case that the relationship between variables is non-linear. A common relationship is one involving *powers* (see Section A.3). Examples are shown in Figs. A.10 and A.11.

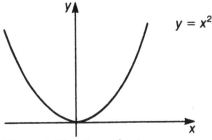

Figure A.10 *A quadratic relationship.*

Figure A.11 *An exponential relationship.*

In cases like the above it is possible to transform the relationships into *linear* relationships by using logarithms.

For example, if $y = x^2$, (A.72)

then taking logarithms of both sides gives $\log y = \log x^2$ (A.73)

(note that it does not matter which base is used)

by applying (A.62), then

$$\log y = 2 \log x. \tag{A.74}$$

This is a *linear* relationship between $\log y$ and $\log x$. The general equation is

$$y = ax^n \tag{A.75}$$

which can be transformed to

$$\log y = \log a + n \log x \tag{A.76}$$

In this way a and n may be calculated on a regression line of $\log y$ on $\log x$.

Another example of this method applies to

$$y = be^{mx} \tag{A.77}$$

Taking logarithms of both sides gives

$\log_e y = \log_e be^{mx}$	(using base e)	(A78)
so $\log_e y = \log_e b + \log_e e^{mx}$	(from A.61)	(A79)
$= \log_e b + mx$	(as e^x is equivalent to the antilog of x and so $\log_e e^x = x$)	(A80)

This is also a linear equation: this time between $\log y$ and x. A regression line can be derived of $\log y$ on x.

A.12   Matrices

A matrix is an ordered array of numbers used to summarize information. This compactness allows the information to be manipulated algebraically. The resulting answers may then be 'disentangled'. For example, suppose

$$y_1 = a_{11}x_1 + a_{12}x_2 \tag{A.81}$$

and

$$y_2 = a_{21}x_2 + a_{22}x_2 \tag{A.82}$$

This may be summarized as

$$[y_1 \quad y_2] = \begin{bmatrix} a_{11} & a_{12} \\ a_{21} & a_{22} \end{bmatrix} \begin{bmatrix} x_1 \\ x_2 \end{bmatrix} \tag{A.83}$$

or $y = A.x$ $\tag{A.84}$

(The notation will soon become clear.)

In this case x is a *column vector*, y is a *row vector* and A is a matrix (in fact a 2×2 matrix — two rows and two columns). A vector is merely a matrix with only one row or one column. Vectors and matrices are no more than convenient ways to store and manipulate information.

Matrices may be added, subtracted and multiplied. And an operation corresponding to division may be applied.

Addition and subtraction of matrices are simple and may be carried out only if they have the same *dimension*, that is the same number of rows and columns.

Matrix A below is a 2×3 matrix, as is matrix B:

$$A = \begin{bmatrix} a_{11} & a_{12} & a_{13} \\ a_{21} & a_{22} & a_{23} \end{bmatrix} \qquad B = \begin{bmatrix} b_{11} & b_{12} & b_{13} \\ b_{21} & b_{22} & b_{23} \end{bmatrix}$$

To add matrices, the equivalent elements are added. Thus:

$$A + B = \begin{bmatrix} a_{11} + b_{11} & a_{12} + b_{12} & a_{13} + b_{13} \\ a_{21} + b_{21} & a_{22} + b_{22} & a_{23} + b_{23} \end{bmatrix} \tag{A.85}$$

Subtraction is just as simple:

$$A - B = \begin{bmatrix} a_{11} - b_{11} & a_{12} - b_{12} & a_{13} - b_{13} \\ a_{21} - b_{21} & a_{22} - b_{22} & a_{23} - b_{23} \end{bmatrix} \tag{A.86}$$

As with numbers it is possible to define an additive inverse of a matrix and an identity element for addition. The additive inverse of A is $-A$, where for each a_{ij} in A the equivalent element in $-A$ is $-a_{ij}$. Thus:

$$A + (-A) = (-A) + A = 0 \tag{A.87}$$

where 0 is the identity element for addition and is a matrix consisting only of zeros. Note that

$$\mathbf{A} + \mathbf{0} = \mathbf{0} + \mathbf{A} = \mathbf{A} \tag{A.88}$$

Note also that

$$\mathbf{A} - \mathbf{B} = \mathbf{A} + (-\mathbf{B}) \tag{A.89}$$

Multiplication is rather more elaborate and can be carried out only if the number of rows in one matrix is equal to the number of columns in another. If $\mathbf{A}$ is 2×3 and $\mathbf{C}$ is 3×3, it is possible to calculate the product $\mathbf{AC}$. The number of columns in the first matrix must equal the number of rows in the second. Thus it is not possible to calculate the product $\mathbf{CA}$. (Note, therefore, that in general matrix multiplication does not obey the *commutative law*.)

A simple example will make this clearer. Let

$$\mathbf{A} = \begin{bmatrix} a_{11} & a_{12} & a_{13} \\ a_{21} & a_{22} & a_{23} \end{bmatrix} \quad \text{and} \quad \mathbf{C} = \begin{bmatrix} c_{11} & c_{12} & c_{13} \\ c_{21} & c_{22} & c_{23} \\ c_{31} & c_{32} & c_{33} \end{bmatrix}$$

Then $\mathbf{D} = \mathbf{AC}$ is obtained as follows

$$d_{11} = a_{11}c_{11} + a_{12}c_{21} + a_{13}c_{31} \tag{A.90}$$

In words d_{11} is obtained by multiplying row 1 of $\mathbf{A}$ by column 1 of $\mathbf{C}$. In general d_{ij} is obtained by multiplying the ith row of $\mathbf{A}$ by the jth column of $\mathbf{C}$.

This can be illustrated by a simple numerical example.

$$\mathbf{A} = \begin{bmatrix} 1 & 2 & 1 \\ 3 & 1 & 3 \end{bmatrix}, \quad \mathbf{C} = \begin{bmatrix} 1 & 1 & 1 \\ 2 & 2 & 2 \\ 1 & 1 & 1 \end{bmatrix}$$

$$\mathbf{D} = \mathbf{AC} = \begin{bmatrix} 1 & 2 & 1 \\ 3 & 1 & 3 \end{bmatrix} \begin{bmatrix} 1 & 1 & 1 \\ 2 & 2 & 2 \\ 1 & 1 & 1 \end{bmatrix}$$

$$\quad\quad\quad\quad\quad\quad d_{11} \quad\quad d_{23}$$

$$d_{11} = 1 \times 1 + 2 \times 2 + 1 \times 1 = 6, \text{ etc.}$$

$$\text{So } \mathbf{D} = \begin{bmatrix} 6 & 6 & 6 \\ 8 & 8 & 8 \end{bmatrix}$$

Note that if $\mathbf{A}$ is 2×3 and $\mathbf{C}$ is 3×3, then $\mathbf{D}$ is 2×3. In general, if $\mathbf{A}$ is $m \times n$ and $\mathbf{C}$ is $n \times p$, then $\mathbf{D}$ is $m \times p$.

These rules may seem obscure but they make sense for the manipulation of the stored information. Further explanation is not possible here.

Returning to the original example (A.83)

$$[y_1 \quad y_2] = \begin{bmatrix} a_{11} & a_{12} \\ a_{21} & a_{22} \end{bmatrix} \begin{bmatrix} x_1 \\ x_2 \end{bmatrix}$$

Multiplying out gives

$$(y_1 \quad y_2) = (a_{11}x_1 + a_{12}x_2 \quad a_{21}x_1 + a_{22}x_2) \tag{A.91}$$

so

$$y_1 = a_{11}x_1 + a_{12}x_2 \tag{A.92}$$

and

$$y_2 = a_{21}x_1 + a_{22}x_2 \tag{A.93}$$

It is also possible to carry out an operation corresponding to division. Division was defined in Appendix A.2 as the multiplication by the multiplicative inverse. Thus if the multiplicative inverse of matrix A is termed A^{-1}, the operation division may be carried out. A^{-1} does not exist unless A is a square matrix (that is, it has the same number of rows as columns) and even then not always, so division requires some care. These points may be illustrated by a simple example. Suppose A, B, C, X and Y are square matrices and:

$$AY = BX + CX \tag{A.94}$$

then

$$AY = (B + C) X \tag{A.95}$$

and, provided A^{-1} exists,

$$A^{-1}AY = A^{-1} (B + C) X \tag{A.96}$$

(by multiplying both sides *on the left* by A^{-1}; remember that matrix multiplication is *not* commutative)

So

$$(A^{-1} A) Y = A^{-1} (B + C) X \tag{A.97}$$

and

$$IY = A^{-1} (B + C) X \tag{A.98}$$

so

$$Y = A^{-1} (B + C) X \tag{A.99}$$

These last three lines require some further explanation. Just as in ordinary multiplication, an element multiplied by its multiplicative inverse gives the identity element. Thus:

$$AA^{-1} = A^{-1} A = I \tag{A.100}$$

And any element multiplied by the identity element gives the original element:

$$IA = AI = A \tag{A.101}$$

The identity matrix for multiplication is

$$I = \begin{bmatrix} 1 & 0 \\ 0 & 1 \end{bmatrix} \quad \text{in the } 2 \times 2 \text{ case.}$$

In general, the identity matrix has 1s along the *leading diagonal* (top left to bottom right) and 0s everywhere else. Readers should check that multiplying a square matrix by I gives the original matrix.

The actual calculation of matrix multiplicative inverses is a tedious and complicated process. Other than for 2×2 matrices it is rarely done by hand. Problems may exist, as not all matrices have a multiplicative inverse. The formula for the 2×2 case is as follows:

If

$$A = \begin{bmatrix} a_{11} & a_{12} \\ a_{21} & a_{22} \end{bmatrix}$$

then

$$A^{-1} = \begin{bmatrix} \dfrac{a_{22}}{a_{11}a_{22} - a_{21}a_{12}} & \dfrac{-a_{12}}{a_{11}a_{22} - a_{21}a_{12}} \\ \dfrac{-a_{21}}{a_{11}a_{22} - a_{21}a_{12}} & \dfrac{a_{11}}{a_{11}a_{22} - a_{21}a_{12}} \end{bmatrix} \tag{A.102}$$

The number $(a_{11} a_{12} - a_{21}a_{12})$ is called the determinant of A and is usually written as $|A|$. Readers should multiply out AA^{-1} and $A^{-1}A$ to ensure that the answer in both cases is I. Thus the inverse A^{-1} exists in the 2×2 case if the determinant of A is non-zero.

A.13 Final comment

For the numerate these notes above should have been straightforward, but for the innumerate the ideas presented may still remain largely impenetrable. If so, readers are encouraged to try again. With some effort an understanding and familiarity should develop, and these should assist a fuller understanding of the main text.

Bibliography and references

Chapter 1

AYENI, B. (1979), *Concepts and Techniques in Urban Analysis*, Croom Helm, London.

BATEY, P.W. and BREHENY, M.J. (1978), 'Methods in Strategic Planning: Part I A Descriptive Review; Part II A Prescriptive Review', *Town Planning Review*, **49**, 259–273 and 502–518.

BATTY, M. (1975), 'In Defence of Urban Modelling', *Journal of the Royal Town Planning Institute*, **61**, 184–187.

BATTY, M. (1976), *Urban Modelling: Algorithms, Calibrations, Predictions*, Cambridge University Press, Cambridge.

BATTY, M. (1979), 'Progress, Success and Failure in Urban Modelling', *Environment and Planning A*, **11**, 863–878.

BAXTER, R.S. (1976), *Computers and Statistical Techniques for Planners*, Methuen, London.

BROADBENT, T.A. (1969), 'Some Techniques for Regional Economic Analysis', *Centre for Environmental Studies Working Paper* No. 61, CES, London.

CATANESE, A.J. (1972), *Scientific Methods of Urban Analysis*, Leonard Hill, London.

CHADWICK, G.F. (1971), *A Systems View of Planning*, Pergamon, Oxford.

DAVIDOFF, P. and REINER, T.A. (1962), 'A Choice Theory of Planning', *Journal of the American Institute of Planners*, **28**, 103–115.

DEPARTMENT OF THE ENVIRONMENT (1973), *Using Predictive Models for Structure Plans*, HMSO, London.

FIELD, B.G. (1984), 'Theory in Practice: the Anatomy of a Borough Plan', *Planning Outlook*, **27**, 68—78.

FOOT, D. (1981), *Operational Urban Models*, Methuen, London.

FRIEND, J.K. and JESSOP, W.N. (1969), *Local Government and Strategic Choice: An Operational Research Approach to the Processes of Public Planning*, Tavistock Publications, London.

GEDDES, P. (1915), *Cities in Evolution*, Ernest Benn, London.

HARRIS, B. (1960), 'Plan or Projection: An Examination of the Use of Models in Planning', *Journal of the American Institute of Planners*, **26**, 265–272.

HARRIS, B. (1964), 'The Uses of Theory in the Simulation of Urban Phenomena', *Journal of the American Institute of Planners*, **30**, 317–322.

HARRIS, B. (1976), 'The Limits of Science and Humanism in Planning', *Journal of the American Institute of Planners*, **33**, 324–335.

ISARD, W. (1960), *Methods of Regional Analysis*, MIT Press, Cambridge, Mass.

KRUECKEBERG, D.A. and SILVERS, A.L. (1974), *Urban Planning Analysis. Methods and Models*, John Wiley, New York.

LEE, C. (1973), *Models in Planning*, Pergamon Press, Oxford.

LINDBLOM, C.E. (1959), 'The Science of Muddling Through', *Public Administration Review*, **19**, 79–88.

McLOUGHLIN, J.B. (1965), 'Notes on the Nature of Physical Planning: Towards a View of Physical Planning', *Journal of the Royal Town Planning Institute*, **51**, 399.

McLOUGHLIN, J.B. (1969), *Urban and Regional Science: A Systems Approach*, Faber and Faber, London.

McLOUGHLIN, J.B. (1985), The Systems Approach to Planning: A Critique', *Centre for Urban Studies and Urban Planning Working Paper* No. 1, University of Hong Kong.

MINISTRY OF HOUSING AND LOCAL GOVERNMENT (MHLG) (1970), *Development Plans: A Manual on Form and Content*, HMSO, London.

PLANNING ADVISORY GROUP (PAG) (1965), *The Future of Development Plans*, HMSO, London.

ROBERTS, M. (1974), *An Introduction to Town Planning Techniques*, Hutchinson, London.

SAYER, R.A. (1976), 'A Critique of Urban Modelling', *Progress in Planning*, **6**(3), 187–254.

WADE, B.F. (1971), 'Some Factors Affecting the Use of New Techniques in Planning Agencies', *Environment and Planning*, **3**, 109–113.

WILSON, A.G. (1974), *Urban and Regional Models in Geography and Planning*, John Wiley, London.

Chapter 2

ADELMAN, I. (1963), 'An Econometric Analysis of Population Growth', *American Economic Review*, **53**, 314–339.

AKERS, D.S. and SIEGEL, J.S. (1969), 'Some Aspects of the Use of Birth Expectations Data from Sample Surveys for Population Projections', *Demography*, **6**(2), 101–115.

ATCHLEY, R.C. (1968), 'A Short-Cut Method for Estimating the Population of Metropolitan Areas', *Journal of the American Institute for Planners*, **34**(4), 259–262.

BAXTER, R. and WILLIAMS, I. (1978), 'Population forecasting and uncertainty at the local and national scale', *Progress in Planning*, **9**(1).

BENJAMIN, B. (1969), 'The Population Census', *Occasional Paper 2, GLC Research and Intelligence Unit.*

BOUVIER, L.F. (1971), 'Estimating Post-Censal Populations of Counties', *Journal of the American Institute of Planners*, **37**(1), 45–46.

BRASS, W. (1974), 'Perspectives in Population Prediction: Illustrated by the Statistics of England and Wales', *Journal of the Royal Statistical Society*, **137**(4), 532–571.

BUXTON, M. and CRAVEN, E. (1976), 'The Policy Significance of Uncertain Demographic Change', in Buxton, M. and Craven, E. (eds.), *The Uncertain Future*, Centre for Studies in Social Policy, London.

COX, P.R. (1976), *Demography*, Cambridge University Press, Cambridge.

DORN, H.F. (1950), 'Pitfalls in Population Forecasts and Projections', *Journal of the American Statistical Association*, **45**, 311–334.

GRAMPIAN REGIONAL COUNCIL (1979), 'Population Forecasting Models', *Research Paper*, No. 2.

GRAMPIAN REGIONAL COUNCIL (1984), *Strategic Forecast — 1984 Update*, Department of Physical Planning, Grampian Regional Council.

GREENBERG, M.R. (1972), 'A Test of Combinations of Models for Projecting the Population of Minor Civil Divisions', *Economic Geography*, **48**, 179–188.

GREENBERG, M.R., KRUECKEBERG, D.A., and MAUTNER, R. (1973), *Long-Range Population Projections for Minor Civil Divisions: Computer Programs and Users' Manual*, New Brunswick, New Jersey: Centre for Urban Policy Research.

HAJNAL, J. (1955), 'The Prospects for Population Forecasts', *Journal of the American Statistical Association*, **50**, 309–322.

HEIDE, H. TER (1963), 'Migration Models and their Significance for Population Forecasts', *The Milbank Memorial Fund Quarterly*, **41**, 56–76.

ISARD, W. (1960), *Methods of Regional Analysis*, Chapters 2 & 3, MIT Press, Cambridge, Mass.

KEYFITZ, N. (1964), 'Matrix Multiplication as a Technique of Population Analysis', *Milbank Memorial Fund Quarterly*, **42**, 68–83.

KEYFITZ, N. (1972), 'On Future Population', *Journal of the American Statistical Association*, **67**, 347–363.

KRUECKEBERG, D.A. and SILVERS, A.C. (1974), *Urban Planning Analysis: Methods and Models*, John Wiley, New York.

LESLIE, P.H. (1948), 'Some Further Notes on the Use of Matrices in Population Mathematics', *Biometrika*, **35** (III & IV), 213–245.

McFARLAND, D.D. (1969), 'On the Theory of Stable Populations: A New and Elementary Proof of the Theorems Under Weaker Assumptions', *Demography*, **6**(3), 301–322.

McKEOWN, T., RECORD, R.G. and TURNER, R.D. (1975), 'An Interpretation of the Decline of Mortality in England and Wales during the Twentieth Century', *Population Studies*, **29**, 391–422.

MASSER, I. (1972) *Analytical Models for Urban and Regional Planning*, David & Charles, Newton Abbot.

MAXWELL, D.E., PHILLIPS, L. and VOTEY, H.L. (1969), 'A Synthesis of the Economic and Demographic Models of Fertility: An Econometric Test', *Review of Economics and Statistics*, **3**, 298–308.

MORRISON, P.A. (1971), *Demographic Information for Cities: A Manual for Estimating and Projecting Local Population Characteristics*, The Rand Corporation, Santa Monica, California.

PARIS, J.D. (1970), 'Regional/Structural Analysis of Population Changes', *Regional Studies*, **4**, 425–443.

POST, A.R. (1969), 'Mobility Analysis', *Journal of the American Institute of Planners*, **35**(6), 417–421.

ROBERTS, M. (1974), *An Introduction to Town Planning Techniques*, Hutchinson, London.

ROGERS, A. (1966), 'Matrix Methods of Population Analysis', *American Institute of Planners Journal*, **32**(1) 40–44.

SCHMITT, R.C. (1952), 'Short-Cut Methods of Estimating County Population', *American Statistical Association Journal*, **47**, 232–238.

SCOTTISH DEVELOPMENT DEPARTMENT (1975), 'Demographic Analysis for Planning Purposes — A manual on sources and techniques', *Planning Advice Note* No. 8.

SIEGEL, J.S. (1972), 'Development and Accuracy of Projections of Population and Households in the United States', *Demography*, **9**(1), 51–68.

SIMPSON, B.J. (1985), *Quantitative Methods for Planning and Urban Studies*, Gower, Aldershot.

WHELPTON, P.K. (1936), 'An Empirical Method of Calculating Future Population', *Journal of the American Statistical Association*, **31**, 457–473.

WILLIAMS, I. (1977), 'Some Sources of Uncertainty Underlying Population Forecasts', *Transactions of the Martin Centre for Architectural and Urban Studies*, **7**, 253–293.

Chapter 3

ASSOCIATION OF METROPOLITAN AUTHORITIES (1985), *Land for private housebuilding*, AMA, London.

BARRAS, R. and BROADBENT, T.A. (1981), 'A Review of Operational Methods in Structure Planning', *Progress in Planning*, **17** (2 and 3).

BLINCOE, W. (1979), 'Analytical Models for Planning Housing Provision: A Critical Assessment of Current Practice', *Department of Civic Design Working Paper* No. 9, University of Liverpool, Liverpool.

CHARLES, S. (1977), *Housing Economics*, Macmillan, London.

COOPERS and LYBRAND (1985), *Land use planning and the housing market: summary report*, HMSO, London.

CULLINGWORTH, J.B. (1969), 'Housing Analysis', in Orr, S.C. and Cullingworth, J.B. (eds.), *Regional and Urban Studies*, George Allen and Unwin, London.

CULLINGWORTH, J.B. (1979), *Essays on Housing Policy*, George Allen and Unwin, London.

DIBLE, J.K. GRANT, R.A. and RANDALL, J.N. (1976), 'Alternative Approaches to Assessing Housing Need', *Housing Monthly*, June, 18–22.

DIBLE, J.K., GRANT, R.A. and RANDALL, J.N. (1976), 'Evaluating the Quality of Housing Stock', *Housing Monthly*, August, 12–18.

DIBLE, J.K., GRANT, R.A., and RANDALL, J.N. (1976), 'Monitoring Trends in the Local Housing Situation', *Housing Monthly*, October, 18–22.

DIBLE, J.K., GRANT, R.A., and RANDALL, J.N. (1976), 'Developing a Housing Strategy', *Housing Monthly*, December, 25–28.

DONNISON, D.V. (1964), *The Government of Housing*, Penguin, London.

ECONOMIC COMMISSION FOR EUROPE (1973), *Housing Requirements and Demand: Current Methods of Assessment and Problems of Estimation*, ECE, Geneva.

HOLMANS, A.E. (1970), 'A Forecast of Effective Demand for Housing in Great Britain in the 1970's', *Social Trends*, 1, 33–42.

HOOPER, A. (1979), 'Land Availability', *Journal of Planning and Environment Law*, 752–756.

HOOPER, A. (1980), 'Land for Private Housing', *Journal of Planning and Environment Law*, 795–806.

HOOPER, A. (1982), 'Land Availability in South-East England', *Journal of Planning and Environment Law*, 555–560.

HOOPER, A. (1985), 'Land Availability Studies and Private Housebuilding', in Barrett, S.M. and Healey, P. (eds.), *Land Policy: problems and alternatives*, Gower, Aldershot (pp. 106–126).

HUMBER, J.R. (1980), 'Land availability — another view', *Journal of Planning and Environment Law*, 19–23.

MacLENNAN, D. (1982), *Housing Economics*, Longman, London.

MORETON, C.G.N. and TATE, T.C. (1973), 'The Vacancy Reserve', *Housing*, 18(6), 18–24.

NEEDLEMAN, L. (1965), *The Economics of Housing*, Staples Press, London.

NINER, P. (1976), 'A Review of Approaches to Estimating Housing Needs', *Centre for Urban and Regional Studies Working Paper*, No. 41, University of Birmingham.

ODLING-SMEE, J. (1975), *The Demand for Housing*, Centre for Environmental Studies, London.

PEIDA (1985), *Housing Demand in Scotland*, PEIDA, Edinburgh.

ROBINSON, R. (1979), *Housing Economics and Public Policy*, Macmillan, London.

ROBSON, B.T. and BRADFORD, M.G. (1984), *Urban Change in Greater Manchester: Demographic and Household Change, 1971–1981*. Report commissioned by the Greater Manchester Council.

SCOTTISH DEVELOPMENT DEPARTMENT (1977), *Scottish Housing Handbook — Assessing Housing Needs: a Manual of Guidance*, HMSO.

SCOTTISH DEVELOPMENT DEPARTMENT (1973), *Circular No. 21/1983*: Private House Building Land Supply: Joint Venture Schemes.

SOCIETY OF SCOTTISH DIRECTORS OF PLANNING/CONVENTION OF SCOTTISH LOCAL AUTHORITIES (SSDP/COSLA) (1984), *Private Sector Housing Methodology Paper*.

STRATHCLYDE REGIONAL COUNCIL (1985), *Population, Household and Housing Projections (Main Report)*, SRC Chief Executive's Department, Glasgow.

STRATHCLYDE REGIONAL COUNCIL (1986), *Population, Household and Housing Projections (Technical Notes)*, SRC Chief Executive's Department, Glasgow.

Chapter 4

ALEXANDER, J.W. (1954), 'The Basic-Nonbasic Concepts of Urban Economic Functions', *Economic Geography*, 30, 246–261.

ANDREWS, R.B. (1956), 'Mechanics of the Urban Economic Base', series of articles in *Land Economics*, 29–31, May 1953–February 1956.

ANDREWS, R.B. (1958), 'Comments re Criticisms of the Economic Base Theory', *Journal of the American Institute of Planners*, 24, 37–40.

AYENI, B. (1979), 'Urban Economic Activity', Chapter 3, in *Concepts and Techniques in Urban Analysis*, Croom Helm, London.

BELL, F.W. (1967), 'An Econometric Forecasting Model for a Region', *Journal of Regional Science*, 7(2), 109–127.

BENDAVID-VAL, A. (1983), *Regional and Local Economic Analysis for Practitioners*, Praeger, New York.

BLUMENFIELD, H. (1955), 'The Economic Base of the Metropolis', *Journal of the American Institute of Planners*, 21, 114–132.

BROWN, H.J. (1969), 'Shift and Share Projections of Regional Economic Growth: An Empirical Test', *Journal of Regional Science*, 9(1), 1–18.

BUCK, T.W. (1970), 'Shift and Share Analysis: A Guide to Regional Policy', *Regional Studies*, 4, 445–450.

CAMBRIDGE ECONOMIC POLICY REVIEW (1982), *Employment Problems in the Cities and Regions of the U.K.: Prospects for the 1980s*, Gower, Aldershot.

ELIAS, P. and KEOGH, G. (1982), Industrial decline and unemployment in the inner city areas of Great Britain: a review of evidence, *Urban Studies*, 19(1), 1–16.

FIELD, B.G. (1977), The fixed-cost hypothesis and the behaviour of unemployment and unfilled vacancies,' *Department of Town Planning Occasional Paper, OP1/77*, Polytechnic of the South Bank, London.

FIELD, B.G. (1985), Some observations on work sharing, *Planning Outlook*, 28(2), 74–76.

FOTHERGILL, P. and GUDGIN, S. (1982), *Unequal Growth: Urban and Regional Employment Change in the U.K.*, Heinemann, London.

HARRIGAN, F. and McNICOLL, I. (1986), Data use and the simulation of regional input-output matrices, *Environment and Planning A*, 18(8), 1061–1076.

HUGHES, J.T. (1969), 'Employment Projection and Urban Development', Chapter 9 in Cullingworth, J.B. and Orr, S.C. (eds.), *Regional and Urban Studies*, George Allen and Unwin, London.

ISARD, W. (1960), *Methods of Regional Analysis*, MIT Press, Cambridge, Mass.

KIRKBRIDGE, D.J. (1970), 'Employment Demand Projections and Future Land-use Requirements', *Journal of the Royal Town Planning Institute*, 56(6), 213–216.

KRUECKEBERG, D.A., and SILVERS, A.L. (1974), 'Regional Income and Employment Analysis', Chapter 12 in *Urban Planning Analysis: Methods and Models*, John Wiley, New York.

LEONTIEF, W. (1966), *Input-Output Economics*, Oxford University Press, New York.

MARSHALL, J.N. (1982), Linkages between manufacturing industry and business services, *Environment and Planning A*, 14(4), 523–540.

MASSEY, D.B. (1973), 'The Basic-Service Categorisation in Planning', *Regional Studies*, 7, 1–15.

MASSEY, D. and MEEGAN, R. (1982), *The Anatomy of Job Loss*, Methuen, London.

MIERNYK, W.H. (1965), *The Elements of Input-Output Analysis*, Random House, New York.

PARIS, J.D. (1970), 'Regional/Structural Analysis of Population Changes', *Regional Studies*, 4, 425–443.

THE PLANNER (1983), *Planning for Employment Regeneration*, 69(5), Theme Issue.

QUINN, D.J. (1986), Understanding accessibility problems of the unemployed, *The Planner*, 72(1), 25–27.

RICHARDSON, H.W. (1978), *Regional and Urban Economics*, Penguin, Harmondsworth.

SAYER, R.A. (1976), 'The Economic Base Model', Section 2 in 'A Critique of Urban Modelling', *Progress in Planning*, **6**(3), 195–201.

SCOTTISH DEVELOPMENT DEPARTMENT (1975), 'Forecasting Employment for Regional Reports and Structure Plans', *Planning Advice Note* No. 4.

STILWELL, F.J.B. (1970), 'Further Thoughts on the Shift and Share Approach', *Regional Studies*, **4**, 451–458.

THORNE, M.F. (1969),'Regional Input-Output Analysis', Chapter 5 in Cullingworth, J.B. and Orr, S.C. (eds.), *Regional and Urban Studies*, George Allen and Unwin, London.

TIEBOUT, C.M. (1962), *The Community Economic Base Study*, Supplementary Paper No. 16, Committee for Economic Development, New York.

ULLMAN, E.L. and DACEY, M. (1960), 'The Minimum Requirements Approach to Urban Economic Base', *Lund Series in Geography*, Series B, **24**, 121–143.

WHIPPLE, R.T.M. (1966), 'Regional Differentials and Economic Planning', *Australian Planning Institute Journal*, **4**, 180–187.

Chapter 5

ALCALY, R.E. (1967), 'Aggregation and Gravity Models: Some Empirical Evidence', *Journal of Regional Science*, **7**(1), 61–73.

ARNOTT, C. and WILLIAMS, J. (1977), 'A New Look at Retail Forecasts', *The Planner*, **63**(6), 170–172.

BEVIS, H.W. and CARROLL, J.D. JNR. (1957), 'Predicting Local Travel in Urban Regions', *Papers and Proceedings of the Regional Science Association*, **III**, 183–197.

BLACK, J. (1966), 'Some Retail Sales Models', Paper presented to the Urban Studies Conference.

BRUNNER, J.A. and MASON, J.L. (1968), 'The Influence of Driving Time upon Shopping Centre Preference', *Journal of Marketing*, 32, 57–61.

BURT, S., DAWSON, J., and SPARKS, L. (1983), Structure plans and retail policies, *The Planner*, **69**(1), 11–13.

CARROTHERS, G.A.P. (1956), 'An Historical Review of the Gravity and Potential Concepts of Human Interaction', *Journal of the American Institute of Planners*, **22**, 94–102.

CASEY, H.J. (1955) 'The Law of Retail Planning Applied to Traffic Engineering'. *Traffic Quarterly*, **9**, 313–321.

COATES, B.E. and RAWSTRON, E.M. (1966), 'Regional Variations in Incomes', *Westminster Bank Review*, February, 28–46.

CORDEY-HAYES, M. and WILSON, A.G. (1971), 'Spatial Interaction', *Socio-Economic Planning Sciences*, **5**, 73–95.

COX, W.E. JNR. (1968), 'The Estimation of Incomes and Expenditure in British Towns', *Applied Statistics*, **17**(3), 252–259.

DAWSON, J. (1980), *Retail Geography*, Croom Helm, London.

DAWSON, J. (1983), Retail impact studies, *The Planner*, **69**(1), 25.

DUNN, E.S. (1956), 'The Market Potential Concept and the Analysis of Location', *Papers of the Regional Science Association*, **2**, 183–194.

FENWICK, I. (1978), *Techniques in Store Location: A Review and Application*, Retail Planning Associates, Northumberland.

GUY, C. (1980), *Retail Location and Retail Planning in Britain*, Gower, Aldershot.

HANSEN, W.G. and LARSHAMANAN, T.R. (1965), 'A Retail Market Potential Model', *Journal of the American Institute of Planners*, **31**, 134–143.

HUFF, D.L. (1963), 'A Probabilistic Analysis of Shopping Centre Trade Areas', *Land Economics*, **39**, 81–90.

HUFF, D.L. (1964), 'Defining and Estimating a Trading Area', *Journal of Marketing*, **28**, 34–38.

ISARD, W. (1954), 'Location Theory and Trade Theory: A Short-run Analysis', *Quarterly Journal of Economics*, **68**, 305–320.

ISARD, W. (1960), 'Gravity, Potential, and Spatial Interaction Models', Chapter 11 in *Methods of Regional Analysis*, MIT Press, Cambridge, Mass.

JONES, C.S. (1969), *Regional Shopping Centres: Their Location, Planning and Design*, Business Books Ltd., London.

KRUECKEBERG, D.A. and SILVERS, A.L. (1974), 'Location and Travel Behaviour', Chapter 9 in *Urban Planning Analysis: Methods and Models*, John Wiley, New York.

LARSHMANAN, T.R. and VOORHEES, A.M. (1966), 'A Market Potential Model and its Application to Planning Regional Shopping Centres', American Marketing Association Conference Proceedings, Autumn, 831–845.

NEDO (1970), *Urban Models in Shopping Studies*, National Economic Development Office, London.

OLSSON, G. (1965), *Distance and Human Interaction: A Review and Bibliography*, Regional Science Research Institute, Pennsylvania.

REILLY, W.J. (1929), 'Methods for the Study of Retail Relationships', *University of Texas Bulletin*, 2944.

RORU (1975), 'Town Planning for Retailing: A Review of Local Authority Policies', *Symposium Proceedings*, Retail Outlets Research Unit, Manchester Business School.

SAYER, R.A. (1976), A Critique of Urban Modelling, *Progress in Planning*, **6**(3), 187–254.

STEWART, J.Q. (1947), 'Empirical Mathematical Rules Concerning the Distribution and Equilibrium of Population', *Geographical Review*, **37**(3), 461–485.

THORPE, D. (ed.) (1974), *Research into Retailing and Distribution*, Saxon House/Lexington Books.

THORPE, D. (1975), 'Assessing the Need for Shops: Or Can Planners Plan?' Planning and Transport Research and Computation Co. Ltd. (PTRC), Summer Meeting, *Proceedings of Seminar M*, 43–52.

URPI (1976), 'Retail Turnover to Floorspace Ratios', *Information Brief* 76/7, Unit for Retail Planning Information, Reading.

WADE, B. (1983), Retail planning without data, *The Planner*, **69**(1), 26–28.

WILSON, A.G. (1970) *Entropy in Urban and Regional Modelling*, Pion, London.

WILSON, A.G. (1974) *Urban and Regional Models in Geography and Planning*, John Wiley and Sons, London.

ZIPT, G.K. (1949), *Human Behaviour and the Principle of Least Effort*, Addison-Wesley, Cambridge, Mass.

Chapter 6

ALLANSON, E.W. (1982), *Car Ownership Forecasting*, Gordon and Breach Science Publishing, London.

ALONSON, W. (1964), *Location and Land-Use*, Harvard University Press, Cambridge, Mass.

ANAS, A. (1984), 'Discrete Choice Theory, Information Theory and the Multinominal Logit and Gravity Models', *Transportation Research*, **17**B(1), 13–23.

BATES, J., ROBERTS, M., LOWES, S. and RICHARDS, P. (1981), *The Factors Affecting Household Car Ownership*, Gower, Aldershot.

BLUNDEN, W.R. and BLACK, J.A. (1984), *The Land-Use/Transport System*, Pergamon, Oxford.

BOYCE, D.E. (1986), 'Integration of Supply and Demand Models in Transportation and Location: Problem Formulations and Research Question', *Environment and Planning A*, **18**(4), 485–489.

BRUTON, M. (1970), *Introduction to Transportation Planning*, Hutchinson, London.

BUTTON, K.J. (1977), *The Economics of Urban Transport*, Saxon House, Farnborough.

CREIGHTON, R.L. (1970), *Urban Transportation Planning*, University of Illinois Press, Illinois.

DEPARTMENT OF THE ENVIRONMENT (1973), *Roads in Urban Areas*, Department of the Environment, HMSO, London.

DEPARTMENT OF TRANSPORT (1972), *Getting the Best Roads for Our Money: The COBA Method of Appraisal*, Department of Transport, London.

DOMENCICH, T.A. and McFADDEN, D. (1975), *Urban Travel Demand: A Behavioural Approach*, North Holland, Amsterdam.

EASTMAN, C.R. (1980), 'A Review of Freight Modelling Procedures', *Transportation Planning and Technology*, **6**, 159–168.

FERTAL, M.O., WEINER, E., BALEK, A.J. and SEVIN, A.F. (1966), *Modal Split: Documentation of Nine Methods for Estimating Transit Usage*, Department of Transportation, Washington D.C.

FINNEY, N.D. (1972), 'Trip Distribution Models', in A.J. Catanese (ed.), *New Perspectives in Urban Transportation Research*, Lexington Books, Lexington, Mass.

FISHER, M.M. and NIJHAMP, P. (1985b), 'Developments in Explanatory Discrete Spatial Data and Choice Analysis', *Progress in Human Geography*, **9**(4), 515–555.

FOOT, D. (1981), *Operational Urban Models*, Methuen, London.

FOWKES, A.S. and BUTTON, K.J. (1977), 'An Evaluation of Car Ownership Forecasting Techniques', *International Journal of Transport Economics*, **4**, 115–143.

GOLDING, S. (1972), 'A Category Analysis Approach to Trip Generation', *Proceedings of the Australian Road Research Board*, **6**(2), 306–324.

HANSEN, W.G. (1959), 'How Accessibility Shapes Land-Use', *Journal of the American Institute of Planners*, **25**, 73–76.

HATSOUKIS, E. (1986), 'Road Traffic Assignment — A Review. Part I: Non-Equilibrium Methods', *Transport Planning and Technology*, **11**(1), 69–79.

HENSHER, D.A. and JOHNSON, L.W. (1981), *Applied Discrete-Choice Modelling*, Croom Helm, London.

JONES, I.S. (1977), *Urban Transport Appraisal*, Macmillan, London.

KRUECKEBERG, D.A. and SILVERS, A.L. (1974), 'Land-Use and Transportation Models', Chapter 10 in *Urban Planning Analysis: Methods and Models*, John Wiley, New York.

LANE, R., POWELL, T.J. and PRESTWOOD-SMITH, P. (1971), *Analytical Transport Planning*, Duckworth, London.

MARTIN, B.V., MEMMOTT, F.W. and BONE, A.J. (1961), *Principles and Techniques of Predicting Future Demand for Urban Transportation*, MIT Press, Cambridge, Mass.

MOORE, E.F. (1957) 'The shortest path through a maze', *Proceedings: International Symposium on the Theory of Switching*, Harvard University, pp. 285–292.

OI, W.Y. and SHULDINER, P.W. (1962), *An Analysis of Urban Travel Demands*, The Transportation Centre, Evanston, Illinois.

O'SULLIVAN, P., HOLTZCLAW, G.D. and BARBER, G. (1979), *Transport Network Planning*, Croom Helm, London.

QUANDT, R. (ed.) (1970), *The Demand for Travel: Theory and Measurement*, Heath Lexington Books, Lexington, Mass.

QUARMBY, D.A. (1967), 'Choice of Travel Mode for the Journey to Work', *Journal of Transport Economics and Policy*, **1**, 273–313.

SASSONE, P.G. and SCHAFFER, W.A. (1978), *Cost-Benefit Analysis: A Handbook*, Academic Press, New York.

STARKIE, D.N.M. (1976), *Transportation Planning, Policy and Analysis*, Pergamon, Oxford.

STOPHER, P.R. and MEYBURG, A.H. (1976), *Urban Transportation Modelling and Planning*, Lexington Books, Lexington, Mass.

VOLMULLER, J. and HAMERSLAG, R. (1984), 'Transportation and Traffic Theory', *Proceedings of the Ninth International Symposium on Transportation and Traffic Theory*, VNU Science Press, Utrecht, Netherlands.

WATSON, P.L. (1974), *The Value of Time: Behavioural Models of Modal Choice*, Lexington Books, Lexington, Mass.

WELLS, G.R. (1975), *Comprehensive Transport Planning*, Charles Griffin, High Wycombe.

WOHL, M. and MARTIN, B.V. (1967), *Traffic Systems Analysis for Engineers and Planners*, McGraw-Hill, New York.

WOOTON, H.J. and PICK, G.W. (1967), 'A Model of Trips Generated by Households', *Journal of Transport Economics and Policy*, **1**, 137–153.

WRIGLEY, N. (1982), 'Quantitative Methods: Developments in Discrete Choice Modelling', *Progress in Human Geography*, **6**(4), 547–562.

ZETTEL, R.M. and CARLL, R.R. (1962), *Summary Review of Major Metropolitan Area Transportation Studies in the United States*, Institute for Traffic and Transportation Studies, Berkeley.

Chapter 7

BOYD-WENNERGREN, E. and NIELSON, D.B. (1970), 'Probability Estimates of Recreation Demands', *Journal of Leisure Research*, **2**(2), 112–122.

BUCHANAN, T., CHRISTENSEN, J.E. and BURDGE, R.J. (1981), 'Social groups and the meanings of outdoor recreation activities', *Journal of Leisure Research*, **13**(3), 254–266.

BURTON, T.L. (1971), *Experiments in Recreation Research*, George Allen & Unwin, London.

BURTON, T.L. (1982), 'A framework for leisure policy research', *Leisure Studies*, **1**, 323–335.

BUTLER, R.W. (1974), 'Problems in the Prediction of Tourist Development; A Theoretical Approach', pp. 49–64 in Matzneter, J. (ed.), *Studies in the Geography of Tourism*, Papers from Conference of the IGU, Working Group, Geography of Tourism and Recreation, Frankfurt/Main.

CHEUNG, H.K. (1972), 'A Day use park visitation model', *Journal of Leisure Research*, **4**(2), 139–156.

CHRISTENSEN, J.E. (1980), 'Rethinking "Social Groups as a basis for assessing participation in selected water activities" ', *Journal of Leisure Research*, **12**(4), 343–356.

CICCHETTI, C.J. (1973), *Forecasting Recreation in the US*, Heath Lexington Books, Lexington, Massachusetts.

CICCHETTI, C.J., SENEA, J.J. and DAVIDSON, P. (1969), *The Demand and Supply of Outdoor Recreation: an econometric analysis*, Bureau of Outdoor Recreation, Washington D.C.

CLAWSON, M. and KNETSCH, J.L. (1966), *Economics of Outdoor Recreation*, Johns Hopkins Press, Baltimore.

COPPOCK, J.T. and DUFFIELD, B.S. (1975), *Recreation in the Countryside: A spatial analysis*, Macmillan, London.

DOTTAVIO, F.D., O'LEARY, J.T. and KOTH, B. (1980), 'The social group variable in recreation participation studies', *Journal of Leisure Research*, **12**(4), 357–367.

ELLIS, J.B. and VAN DOREN, C.S. (1966), 'A Comparative Evaluation of Gravity and System Theory Models for Statewide Recreational Traffic Flows', *Journal of Regional Science*, **6**(2), 57–70.

ELSON, M.J. (1974), 'The uses and assumptions of probability models in recreation planning', *Planning Outlook* (Special Issue), Planning for Recreation, 65–72.

FIELD, D.R. and O'LEARY, J.T. (1973), 'Social groups as a basis for assessing participation in selected water activities', *Journal of Leisure Research*, **5**(2), 16–25.

GRATTON, C. and TAYLOR, P., (1985), *Sport and Recreation — An Economic Analysis*, E. & F.N. Spon, London.

HENRY, I.P.D. (1980), 'Approaches to Recreation Planning and Research in the District Authorities of England and Wales', *Leisure Studies Association Quarterly*.

MANSFIELD, N.W. (1971), 'The Estimation of Benefits from Recreation Sites and the Provision of a New Recreation Facility', *Regional Studies*, **5**(2), 55–69.

NORTH WEST SPORTS COUNCIL (1972), *Leisure in the North West*, NWSC, Salford.

ROBERTS, M. (1974), *An Introduction to Town Planning Techniques*, Hutchinson, London.

SCOTTISH TOURISM AND RECREATION STUDIES (undated), *Planning for Sport, Outdoor Recreation and Tourism*: **2**, A Guide to the preparation of initial regional strategies, Countryside Commission for Scotland.

SEARLE, G.A.C. (ed.) (1975), *Recreational Economics and Analysis*, Longman, Harlow.

SETTLE, J.G. (1977), 'Leisure in the North West: A tool for forecasting', *Sports Council Study*, 11, Sports Council/North West Sports Council.

SHAW, M. (1984), *Sport and Leisure Participation and Lifestyle in Different Residential Neighbourhoods*, Sports Council, ISSRC.

SILLITOE, K.K. (1969), *Planning for Leisure*, HMSO, London.

SMITH, R.J. (1971), 'The evaluation of recreation benefits: The Clawson Method in practice', *Urban Studies*, **8**(2), 89–102.

SMITH, S.L.J. (1983), *Recreation Geography*, Longman, London.

SNEPENGER, D.J. and CROMPTON, J.L. (1984), 'Leisure activity participation models and the level of discourse Theory', *Journal of Leisure Research*, **16**(1), 22–33.

SPORTS COUNCIL (1968), *Planning for Sport — Report of a Working Party on Scales of Provision*, Central Council of Physical Recreation, London.

TORKILDSEN, G. (1983), *Leisure and Recreation Management*, E. & F.N. Spon, London.

VEAL, A.J. (1974), 'Estimating demand at Urban recreation facilities', *Planning Outlook* (Special Issue), Planning for Recreation, 58–64.

VEAL, A.J. (1982), 'Planning for Leisure: Alternative Approaches', *Papers in Leisure Studies*, **5**, Polytechnic of North London, London.

VEAL, A.J. (1984), 'Leisure in England and Wales', *Leisure Studies*, **3**, 221–229.

VICKERMAN, R.W. (1974), 'The evaluation of benefits from recreation projects', *Urban Studies*, **11**, 277–288.

WILKINSON, P.F. (1973), 'The use of models in predicting the consumption of outdoor research', *Journal of Leisure Research*, **5**(3), 34–47.

Chapter 8

BATEY, P.W.J. and BREHENY, M.J. (1978), 'Methods in Strategic Planning, Part I: A Descriptive Review', *Town Planning Review*, **49**(3), 259–273.

BATEY, P.W.J. and BREHENY, M.J. (1978), 'Methods in Strategic Planning, Part II: A Prescriptive Review', *Town Planning Review*, **49**(4), 502–518.

BRACKEN, I. (1982), 'New Directions in Key Activity Forecasting', *Town Planning Review*, **53**(1), 51–64.

BRACKEN, I. and HUME, D. (1981), 'Forecasting Methods and Techniques in Structure Plans: Lessons from the Welsh Plans, *Town Planning Review*, **52**(3), 375–389.

BREHENY, M.J. and ROBERTS, A.J. (1978), 'An Integrated Forecasting System for Structure Planning', *Town Planning Review*, **49**(3), 306–318.

COCKHEAD, P. and MASTERS, R. (1984), 'Forecasting in Grampian: Three Dimensions of Integration', *Town Planning Review*, **55**(4), 473–488.

DEWHURST, R. (1984), 'Forecasting in Greater Manchester: A multi-regional approach', *Town Planning Review*, **55**(4), 453–472.

GRAMPIAN REGIONAL COUNCIL (1978), 'An Integrated Forecasting Methodology for Strategic Planning', *Planning Research Paper*, No. 1, Grampian Regional Council, Aberdeen.

TURNER, K. (1982), 'Integrated forecasting — An End to the Means', *British Urban and Regional Information Systems Association Newsletter*, **55**.

Chapter 9

SAYER, R.A. (1976) A Critique of Urban Modelling, *Progress in Planning*, **6** (3).

Index

Milton Keynes UK
Ingram Content Group UK Ltd.
UKHW031148141024
449569UK00024B/980